The Challenges of Island Studies

Ayano Ginoza
Editor

The Challenges of Island Studies

 Springer

Editor
Ayano Ginoza
Research Institute for Islands
and Sustainability
University of the Ryukyus
Nishihara, Okinawa, Japan

ISBN 978-981-15-6287-7   ISBN 978-981-15-6288-4 (eBook)
https://doi.org/10.1007/978-981-15-6288-4

© Springer Nature Singapore Pte Ltd. 2020
This work is subject to copyright. All rights are reserved by the Publisher, whether the whole or part of the material is concerned, specifically the rights of translation, reprinting, reuse of illustrations, recitation, broadcasting, reproduction on microfilms or in any other physical way, and transmission or information storage and retrieval, electronic adaptation, computer software, or by similar or dissimilar methodology now known or hereafter developed.
The use of general descriptive names, registered names, trademarks, service marks, etc. in this publication does not imply, even in the absence of a specific statement, that such names are exempt from the relevant protective laws and regulations and therefore free for general use.
The publisher, the authors and the editors are safe to assume that the advice and information in this book are believed to be true and accurate at the date of publication. Neither the publisher nor the authors or the editors give a warranty, expressed or implied, with respect to the material contained herein or for any errors or omissions that may have been made. The publisher remains neutral with regard to jurisdictional claims in published maps and institutional affiliations.

This Springer imprint is published by the registered company Springer Nature Singapore Pte Ltd.
The registered company address is: 152 Beach Road, #21-01/04 Gateway East, Singapore 189721, Singapore

Preface

This book is the culminative product of a multi-year research project led by the Research Institute for Islands and Sustainability (hereafter RIIS) at the University of the Ryukyus in Okinawa. The research was titled "Conceptualizing *Tosho chiiki kagaku kenkyu* (Regional Science for Small Islands, or RSSI) toward a *Jiritsu* Model of Island Societies." This project was funded by the Japanese Ministry of Education, Culture, Sports, Science, and Technology. The project was carried out during the 2016–2018 Japanese academic years, involving faculty members in humanities and social sciences from the University of the Ryukyus. After the conclusion of the project, RIIS held an international symposium titled "Prospects and Challenges for Envisioning Regional Science for Small Islands" on March 1, 2020, where three prominent figures in international island studies were invited to join two of the project members to share and discuss key issues in the field of island studies.

This book consists of seven chapters, including six papers and a discussion of current topics in island studies. In the introduction, "An Archipelagic Enunciation from Okinawa Island," I have illuminated key Japanese terminology in the regional science for small islands that is not readily translatable into English words: *jiritsu* (autonomy), *shutai* (subject), and *tojisha* (those primarily affected). By highlighting the issues and usefulness of such keywords through the interdisciplinary lenses of gender and nation, this chapter places RSSI in conversation with other challenges in the field of island studies.

In "Island Studies in (and outside of) the Academy: The State of This Interdisciplinary Field," James E. Randall evaluates trends in the "emerging" institutional structures associated with island studies. He delineates the state of the emerging field from outside and inside the academic field. In doing so, Randall envisions the future of this interdisciplinary field of enquiry.

With a geographical focus on Guåhan/Guam, an unincorporated territory of the United States and heavily militarized by the U.S., Ronni Alexander attempts to redefine security as both "being and feeling safe," in her chapter titled "Islands as Safe Havens: Thinking about Safety and Security in Guåhan/Guam" Taking a feminist perspective, she analyzes interviews and conversations with Chamorro and the other residents on the island. She concludes that, in a place where colonization

v

and militarization reinforce masculinities based on power and the desire for protection through military means, being safe is only attainable in unsafe ways.

Encompassing the U.S. Navy's expansion from the Pacific Islands to the Indo-Pacific, Elizabeth DeLoughrey examines the geopolitics in relation to the field of "critical ocean studies" in her chapter titled "Island Studies and the U.S. Militarism of the Pacific." Bringing the work of Chamorro poet Craig Santos Perez into conversation with critical ocean studies, which link "submarine immersions, multispecies others, feminist and Indigenous epistemologies, and the acidification of an Anthropocene ocean," DeLoughrey argues that this strategic military grammar is vital for understanding island studies. She expounds on ways in which the U.S. Navy has militarized from the Pacific Islands to the Indo-pacific.

In "The Perspective of Cultural Heritage/Cultural Landscape in Critical Island Studies," So Hatano discusses island spaces as "an emergent product of relations" to counter the colonial discourses that define island characteristics in terms of remoteness and isolation. As an example, he uses the Jinguashi Mines, a potential World Heritage Site, in the northern part of Taiwan (Massey 2005).

In "The Possibilities of Phylogenetic Tree Studies in Ryukyuan Languages Research," Shigehisa Karimata borrows from biological classification methods and applies the use of phylogenetic trees in the study of Ryukyuan languages. Karimata touts that building of a framework for systematic and comprehensive study of the Ryukyuan languages necessitates what he calls a "linguistic phylogeography in dialectics" as an approach that benefits from interdisciplinary input that combines archeology, physical anthropology, folklore, and folk music.

The final chapter titled "Prospects for Critical Island Studies" consists of a dialogue among the aforementioned scholars and the audience. The dialogue begins with an introduction to the concept of regional science for small islands (RSSI) by Yoko Fujita, the director of the Research Institute for Islands and Sustainability at the University of the Ryukyus. She explains that RSSI is an interdisciplinary field for studying small islands which aims to explore the challenges and solutions for sustainable and autonomous development of small islands. Further, RSSI takes three major approaches to the issues small islands have: employing normative science (a theoretical approach), empirical science (simulations and a case study approach), and practical science (approaches for implementing studies' findings). Following this, the panelists, Randall, Alexander, DeLoughrey, Hatano, and Karimata, engage with each other's papers as well as answering questions from the audience about ongoing and future challenges in island studies.

On behalf of the Research Institute for Islands and Sustainability, I express appreciation to all the contributors of the articles as well as the participants in the dialogue. My deepest appreciation goes to the editors of Springer for making this edited volume possible.

Nishihara, Japan   Ayano Ginoza

Contents

An Archipelagic Enunciation from Okinawa Island................. 1
Ayano Ginoza

Prospects for Island Studies

Islands as Safe Havens: Thinking About Security and Safety
on Guåhan/Guam 17
Ronni Alexander

Island Studies and the US Militarism of the Pacific................. 29
Elizabeth DeLoughrey

Island Studies Inside (and Outside) of the Academy: The State
of this Interdisciplinary Field................................. 45
James E. Randall

The Perspective of Cultural Heritage/Cultural Landscape in Critical
Island Studies.. 57
So Hatano

The Possibilities of Phylogenetic Tree Studies in Ryukyuan
Languages Research 79
Shigehisa Karimata

Panel Discussions

Prospects for Critical Island Studies 95
Ayano Ginoza, Ronni Alexander, Elizabeth DeLoughrey,
James E. Randall, So Hatano, and Shigehisa Karimata

An Archipelagic Enunciation from Okinawa Island

Ayano Ginoza

As the only center in the areas of humanities and social sciences at the University of the Ryukyus that focuses on islands research, the Research Institute for Islands and Sustainability (hereafter RIIS) has completed three multi-year research projects since the time RIIS was still known as the International Institute for Okinawan Studies. The name change occurred in close association with the University of the Ryukyus' articulation of its situated identity as an island university, stressing its position in Japan's only prefecture consisting of a chain of small islands. The projects were titled: "Creating a New [Field of] Island Studies: The Ryukyu Archipelago as a Hub That Links Japan to East Asia and Oceania" (2011–2015), "Creating Okinawan Gender Studies" (2011–2015), and "Conceptualizing RSSI: Toward the Creation of an Autonomy (*jiritsu*) Model of Island Societies" (2016–2019), which this chapter focuses on. The final project was embodied in two forms: an international symposium held in March 2019, to which the contributors to this anthology were invited, and an anthology publication titled *The Challenges of Regional Science for Small Islands* (*Tosho-chiiki-kagaku toiu chosen*), 2019, edited by two of the project members, the Director of RIIS, Yoko Fujita, and historian Daisuke Ikegami. Among the many challenges the editors contemplated since the first island project "Creating a New [Field of] Island Studies," I elucidate three keywords that appear throughout the course of the research projects—*tojisha* (those primarily affected), *jiritsu* (autonomy), and *shutai* (subject). While each project's goals and objectives varied, these three keywords tethered them to the core mission of the institute.

As a newcomer to the field of island studies, my intention is not to offer an extensive literature review of island studies and to locate RIIS within it. Nor does this chapter reflect the overall institutional vision or scope. Rather, this chapter intends to unpack the keywords of the multi-year research projects to locate the project that embodies the institute's directions and objectives by engaging in a constructive

---

A. Ginoza (✉)
Research Institute for Islands and Sustainability, University of the Ryukyus, Nishihara, Japan
e-mail: ginoza@eve.u-ryukyu.ac.jp

© Springer Nature Singapore Pte Ltd. 2020
A. Ginoza (ed.), *The Challenges of Island Studies*,
https://doi.org/10.1007/978-981-15-6288-4_1

conversation with the field of island studies and Okinawan studies in an interdisciplinary manner—the location where RIIS's mission overlaps. By doing so, I hope to engage with the vision, challenges, and objectives as a faculty member of RIIS through the interdisciplinary lenses of gender and nation upon which my recent research centers. The principal question I have in this chapter is how RSSI, an emergent interdiscipline presently in the process of realization, might establish useful grounds for enriching the current discussions in the field of island studies both domestically and internationally. My proposition is to consider contextualized ways of understanding *island* or *islandness* from the location of Okinawa Island. This is best done by applying Naoki Sakai's theory of translation from Japanophone, which he calls a "heterolingual address." I argue that a mode of heterolingual address allows for useful comparisons with many other locales, situations, and moments, in particular engagements with the contemporary US empire, as well as various instances where the hegemony of English has been internalized in academia, wherein island studies scholars from diverse linguistic communities preform acts of translation without questioning the effect of translation.

Below, drawing from Godfrey Baldacchino, a founder of the *Island Studies Journal* and a most prominent scholar in island studies, I first discuss the key term *hybridity* in the current discussion in the field of island studies, where it resonates the challenges and contributions RSSI aspires to make to the field (Baldacchino 2008). Then, I consider Sakai's theory of a *heterolingual address* to discuss the effects of translation and the role of the translator in producing what he calls a *nonaggregate community* in a manner that echoes Baldacchino's hybridity. Next, I discuss the keywords *shutai*, *jiritsu*, and *tojisha* in related epistemologies in the Japanese academic disciplines where they are employed, mobilized, and problematized in their practices and applications. I conclude this chapter with the prospects for RSSI by exploring a useful application of Judith Butler's theory of gender performativity to islandness.[1]

Hybridity and Heterolingual Address

The exploration of the contours of RSSI begins with a declaration of island hybridity made by Baldacchino. He states that hybridity is the norm of islands:

> The pursuit of nissology, or island studies, calls for a re-centering of focus from mainland to island, away from the discourse of conquest of mainlanders, giving voice and platform for the expression of island narratives. Yet, studying islands 'on their own terms', in spite of its predilection for "authenticity", is fraught with epistemological and methodological difficulties. The insider/outsider distinction does not work all that well when it comes to islands, where hybridity is the norm. (Baldacchino 2008, p. 37)

---

[1] I do not claim to speak for the Research Institute's overall goals and future trajectory in this chapter, nor does the argument made in this paper reflect the mission or objective of RIIS, but rather this is a theoretical attempt at unpacking the key terminologies used in the mission statement.

In his essay, hybridity is listed as one of the keywords along with other keywords: island studies, outsiders, insiders, mainlanders, islanders, and re-centering. These keywords play a central role in studying islands and "identifying epistemological and methodological challenges to the pursuit of island studies" (Baldacchino 2008, p. 44). Hybridity is defined in relation to island identity and location of subjects such as "islanders, natives, settlers, tourists, second-home owners" as well as those who would study them…who may be locals as well as outsiders (mainlanders, continental dwellers)—looking in" (Baldacchino 2008, p. 38). According to Baldacchino, such hybridity of island(er)s constitutes a *nervous duality* and is problematic in nurturing it while addressing their essence. Hence, hybridity impacts, and has paradigmatic effects on, understanding islandness. Drawing from Bhabha (1995), Baldacchino elucidates that hybridity can be "a viable strategy for subverting the narratives and representations promulgated and imposed by external dominant powers and cultures" even if hybridity is produced as "the outcome" of varying degrees of coloniality or internalization of it (Baldacchino, pp. 41–42; Bhabha 1995). Hybridity of language in communication and format is no exception to the island studies "*problematique*," according to Baldacchino, especially when translations between and among multiple languages are treated simply as a bridge for the unified national languages.

The problem of translation, according to Naoki Sakai, illuminates the "social and even political issues involved in translation" (Sakai, p. 2). According to Sakai, the social and political issues involved in translation, or *the regimes of translation*, is brought about as an effect of the *problematique* in the process of "postulate[ing] the two unities of the translating and the translated language as if they were autonomous and closed entities through a certain representation of translation" (Sakai, p. 2). Building on the discussion of how the Japanese language emerged as the national language by separating itself from the Chinese language while simultaneously constituting a difference between the subject/Japanese and the other/non-Japanese, Sakai elaborates that regimes of translation confine us within the discourse regulated to the idea of national language and the schema of its configuration. Such manner of schema of such configuration, which Sakai terms a *homolingual address*, assumes the position of the addresser as the "representative of a putatively homogeneous language society and relates to the general addressees, who are also representative of an equally homogeneous language community" (Sakai 1997, p. 4). He proposes to develop a set of tropes that requires "an overall reconsideration of the basic terms in which we represent to ourselves how our translational enunciation is a practice of erecting or modifying social relations" (Sakai 1997, p. 2).

One of the tropes is the mode of putative collectivity of *we* (us, ourselves), the addressee, to not coincide with a linguistic community built around the assurance of reciprocal apprehension. But as a trope, it becomes an attempt at "reach[ing] out to the addressees without either an assurance of immediate apprehension or an expectation of uniform response from them. In this way, 'we' are rather a 'nonaggregate community' for the addressees would respond to my delivery with varying degrees of comprehension" (Sakai 1997, p. 2, emphasis added). This mode of relating the addresser to the addressee is what he calls a *heterolingual address*. If the field of island studies assumes a hybrid *we* as participants and nonaggregate community of addressees, then *we* have to write through the attitude of a heterolingual address. The trope may be a methodology for erecting or modifying social relations in our transnational enunciation.

In relation to RIIS, another trope is the position and the role of the translator. In heterolingual address, a translator occupies a position where multiplicity of language is implicated, and she performs the role of a heterolingual agent of interpretation. Sakai delineates this complex theory comparatively with a twofold mode of homolingual translator. First, the translator's role will be erased in the transactional representation of translation of one language's transfer into an equivalent message in another language. Second, incommensurability exists not between the addresser and the addressee but between one linguistic community and another (Sakai 1997, p. 10). Hence, Sakai calls into question the supposed discernibility of *interlingual* from *intralingual translation*, of translation between separate languages from rewording within the same language unity, and claims that not only the professional translator, but the rest of us as well, are responsible for the task of the translator (Sakai 1997, p. 10).

If I consciously commit to a heterolingual address, and I am aware of the sociality a translation may produce, then my heterolingual address includes consciously performing the role of translator to transform the discontinuity of the addresser (the original subject of enunciation) with the addressee (the reader). In doing so, a translator performs a translation of the untranslatable as a testimony to the sociality of the translator and "exposes the presence of a nonaggregate community between the addresser and the addressee, as to the translatable itself" (Sakai 1997, p. 14).

This understanding may provide a critical methodology for RSSI, where a decolonial future is imagined and being created, a reality of hybridity of language, erasure, loss, and relearning of languages, and the problem of representation of islanders in the translation may demand more serious unpacking. The critical role of the translator, which many islander scholars perform daily, becomes even more evident because in a profound and interweaving manner, indigenous scholars from islands—including myself—must, when we preform both spoken and unspoken translations through publications, talks, and teaching, disturb the nomocracy of homolingual address. We must eschew translation as the simplified transfer of a message from one national language to the other. Heterolingual address presents a nonaggregate community as it acknowledges the positionality of the translator's performance during translation, whereas a homolingual address places a translation as a form of communication between two fully formed different language groups.

In homolingual address, one side makes the other visible by giving rise to a type of subject formation. Furthermore, Sakai reminds us that such homolingual address served to reify and essentialize the Japanese culture and nation in Japanese studies in the United States and Europe as in Japan today, wherein looms an equally essentializing uniqueness of the West:

> Only where it is impossible to assume that one should automatically be able to say what one oneself means and an other able to incept what one wants to say—that is, only where an enunciation and its inception are, respectively, a translation and a counter translation—can we claim to participate in a nonaggregate community where what I want to call the

heterolingual address is the rule, where it is imperative to evade the homolingual address. (Sakai, 1997, p. 7)

Creating a nonaggregate community may have the potential to facilitate a critical engagement in island matters and to help foster the study of island plurality where hybridity is the acknowledged norm. When creating an RSSI that aims to encompass domestic and translational studies of island regions, it will be prudent to be conscious of the danger of reinforcing the colonialities and essentializing culturalism in its translations of scholarship. The proposed concept of heterolingual translation does not only refer to the realm of linguistics and literature, but it also shares much with postcolonial commitment to rethinking imperialism. This approach has the potential for resistance for islanders who are defined in the relativism of the continents and their accompanying empires. In the following section, I discuss an attitude of heterolingual address in RIIS's inaugural commitment to an *island* turn.

RSSI and the Practice of Heterolingual Address

The RIIS at the University of the Ryukyus was established in 2009 and marked the transition from the previous name, the International Institute for Okinawan Studies. In "From the Director: Regional Science for Small Islands from Okinawa," the director of RIIS, Yoko Fujita, inaugurates the beginning of RSSI, a new field of island studies, capturing the moment of *island turn* at the University of the Ryukyus:

> Since its establishment in 2009, the International Institute for Okinawan Studies developed a diverse range of research themes by promoting international and interdisciplinary research regarding Okinawa and other regions in close relationships with Okinawa. In the process, we have gained an understanding that the issues shared between Okinawa and other island regions derive from aspects of islandness. Such an understanding has enabled us to encompass research themes to include islands both domestic and abroad that share Okinawa's and the region's challenges. In order to further develop these achievements, and to strengthen our network with researchers, universities, and research organizations who are tackling both domestic and international island studies themes, with the goal of further establishing and developing the academic field of "RSSI," the name of the institute has changed to the Research Institute for Islands and Sustainability (RIIS).
>
> At the new research institute, we will attempt to develop a "regional science for small islands" through three scientific approaches—the normative, empirical, and practical. Research will center on topics such as the diversity of islands, cultures, and communities, the relationship with islands across the oceans, and the social and economic systems suitable for islands, with an objective of promoting research that facilitates theoretical, empirical, and practical considerations. Commencing with Okinawa, we aim to contribute to the autonomous (*jiritsuteki*) and sustainable development of island regions. (Fujita 2018a)

What compelled the process of the name change is an acknowledgement of the interrelatedness of island matters that closely resonates with international Okinawan studies, yet we are cognizant that Okinawan matters cannot always be contained within the framework of identities of Okinawan studies. Because of its already plural articulations of Okinawa, RIIS took a courageous and challenging approach to the matter by building RSSI theories and methods. If my interpretation has any validity, granted that I am not representing or speaking for the institute's overall ideology, RSSI can be considered an attitude for enabling a nonaggregate community. This new sensibility cannot be elucidated without an interisland lens. In this sense, the name change to the RIIS as an institutional *island turn* is not to disengage from international Okinawan studies but rather is grounded in the moment of realization that Okinawan studies means more than regional and area studies and could be a site for emerging contours of critical methodology and solidarity as Ryukyuans decolonize the intersections of small islands and continents/larger nation-states within and beyond Okinawa.[2] In a way, RIIS's effort can be understood as a necessity realized by some scholars at the University of the Ryukyus for an interventional solidarity of island scholarship but still maintaining its place on the Ryukyu-Okinawa Archipelago. This effort was compiled in an anthology, *Tosho-chiiki-kagaku toiu chosen* (*The Challenges of RSSI*, 2019), co-edited by Daisuke Ikegami and Yoko Fujita. Ikegami and Fujita (2019) state that the important strategy is to systematically study islands to underscore how larger continents or nation-states have tactically crafted relativist conceptualization of islands by their remoteness, marginality, smallness, and vulnerability to larger continents.

To consciously or unconsciously accept the given characteristics of islands, or islandness, by their comparative size and distance to the continents may not only immobilize island studies scholarship, but it may also risk islandness being reduced to a framework of *dis*ability. It may even cast islands in the position of a *dis*abled other, as victims of an inevitable geopolitical determinization. Hence, the dialectic approach can be maintained as an unwavering hegemonic device. Within that discourse, islanders who comprise the island lives can consequently be seen as disadvantaged and disabled. Rather than accepting such a discourse, it is important to destabilize the dominant narratives of disabled and able-bodied normalcy. As a scholar of disability, Campbell (2009) illuminates ableism as a "construct of social relations where a network of beliefs, processes, and practices that produces a particular self and body that is produced as the perfect, species-typical and therefore essential and fully human, [while disability is defined as] a diminished set of being human" (Campbell 2009, p. 44). Erevelles (2011) elucidates that both are constituted in social relations within transnational capitalism. I argue the establishment of RIIS is indicative of approaches that challenge and destabilize normative epistemology of previous dialectics.

---

[2]The original Japanese word translated as *issue* is *kadai*. Kadai has multiple meanings, such as assignments, tasks, themes, or subjects. Whichever the best definition might be, the implication is something to be solved or tackled. If *islandness* in that context can be understood as characteristics of islands, islands across the world may have the potential to share the issues brought about by being islands. What is important here is not to understand islandness as the source of *kadai* to be solved or accept that islandness means already born with *kadai*, but to contextualize the ways in which islandness is made, by whom and what, and when to deconstruct the discourse of the *issues* that islands may share in common.

The relativist dichotomy inevitably enables a discourse of dependency of *dis*abled small island states on the larger *able*-bodied nation-states for economic, military, and political assistance. This discourse is constructed to work to the advantage of larger nation-states, just as masculinity does not benefit from its hegemonic existence without the feminine "other," nor the Occident without the Orient. In a most deductive manner, the model works when both the dependent and the depended upon coexist in the same realm, in other words, as a power effect, necessitating the larger-states' ontological dependency on their unequal relationship with the small islands. Interdependency and co-dependent relationships between islands and continents are apparent. Yet, these relationships are often hidden in the history and milieu of everyday island life.

Critical of such inscriptive discourse of power relations that has been self-endowed to the larger states, RSSI's work may insist on restoring the *world of islands*, to borrow Baldacchino's term, by "considering islands subjectively (*shutaiteki*)" and maintaining *tojishasei* (elements of people or parties concerned) in order to realize a jiritsuteki model. Thus, these keywords repeatedly serve to contextualize the RIIS framework as a theoretical and methodological underpinning for its studies of small islands (Fujita 2018b).

Tojisha

The concept of *tojisha* has been developed in the field of law, social welfare, psychology, sociology, and gender studies in Japan. The keyword *tojisha* does not have an equivalent word in English, but it is commonly translated into a phrase such as "a person or party concerned; a concerned and an interested party; a participant" (Watanabe et.al. 2017). For instance, in social welfare, *tojisha* is defined as people with disabilities; in law, as a person impacted by an incident or a problem; and in social work as a client of a worker. The former connotates the subject as lacking abilities; the latter constructs a hierarchical relationship between experts and a client in need of the experts' service.

Chizuko Ueno, a Japanese feminist scholar who published several books on *tojisha*, states that the concept was inevitably developed in the field of disability studies to acknowledge and sanction the individual autonomy of people with disabilities who have been marginalized. Ueno defines *tojisha* as people who are deprived of the right to individual autonomy over their own future, with a mission to share, accumulate, verbalize, and theorize the *tojisha* experiences of marginalized people (Ueno 2010, 2011). In *Tojisha Shuken* (sovereignty), Shoji Nakanishi and Chizuko Ueno defined *tojisha* as individuals to whom a need or needs are attributed. Employing an example from the women's liberation movement where women articulated women's issues as social issues, Ueno and Nakanishi argue that people do not already exit as *tojisha*, but those who are faced with problems *become tojisha* by achieving "positional subjectification" (Nakanishi and Ueno 2003, pp. 2–3). Here, Ueno distinguishes *being tojisha* from *becoming tojisha*. In such a differentiation, *tojisha* emerge as active subjects who identify the issues that pertain to them:

> *Tojisha* is not synonymous with people with problems. There would be no needs in a society where people adapt to the society that produces problems. The definition of needs originates in lack and shortages. Individuals will know their needs and become *tojisha* when they identify current situations as lacking their desired state and imagine a better reality. Needs are not in preexistence but created. To create needs is to envision another society. (Ueno 2011, p. 79)

In the process of becoming *tojisha*, individuals become agents with needs. Ueno (2011) explains that *tojisha*, as agents with needs, are therefore subjects who view fulfillment of such needs as their social responsibility. Nakanishi and Ueno (2003) define the studies as a collection of discourses and theories reflecting *tojisha*'s own experiences, a listening to *tojisha* voices with an emphasis on the subjectivity of the participants, thereby distinguishing them from the supposedly objective expertise established by outside authorities. What Ueno and Nakanishi may not have considered is Foucault and Butler's theory of subjects that do not exist a priori but rather are created as an artifact of power. If the binary opposition is an effect of power differences, then, deploying identificatory categories may be used by agents (such as a state) to "conjure and regulate subjects through classificatory schemes, naming and normalizing social behaviors as social positions" (Brown 1993, p. 393).

Despite extensive definitions, what definitions of *tojisha* have in common is the position of social marginality. To me, the literature of *tojisha* studies (*tojisha gaku*) appears insufficient to understand *tojisha* in RSSI. This is because it unwittingly begins from the conditions of social marginalization. In "Purpose and Philosophy" on the RIIS homepage, Fujita states:

> It can be said that, in terms of "*tojisha-sei*," (a Japanese concept describing those who are directly affected by issues, or those who are in such a state) Okinawa, as an island, is a most suitable place for the promotion of the Regional Science for Small Islands. This is because Okinawa, while known for its distinctive characteristics, is also the sole prefecture in Japan made up of only small islands, and shares remarkable commonalities with other islands. This "*tojisha*" perspective does not originate in the viewpoint that defines islands as "periphery" or "remote," but instead is an approach in which we consider the problems of islands subjectively (*shutaiteki*) and empirically from the position of islands themselves. That is to say, our positionality as a "*tojisha*" enables us to grapple with resolving problems facing islands with empathy for other islands. We will utilize these ideas of "*tojisha-sei*" and "empathy" to expand our cooperation with researchers and research organizations in the field and to investigate various island issues with depth and breadth. (Fujita 2018b)

Fujita is careful in the way she uses *tojisha-sei* and not *tojisha* in her remarks. The conscious deployment of the term suggests an implication of deconstructing the marginality of islanders or islands and rejects affirming the position of the marginalized. If the identity of *tojisha* is already a fixed position, then, as Ueno argues, the law must provide them with an autonomy of power. If autonomy is something granted to the marginalized by those in power, then the model is always unable to deconstruct the power relationship since employing it serves to reinforce it, far from the *jiritsu* model of RSSI. Moreover, what RIIS advocates in *tojisha* is to exhibit empathy of care for islanders by researchers who live on the islands. Ikegami and Fujita (2019) state that their project was carried out with empathy for the islands and as inhabitants of islands (*tojisha-sei*).

In bridging the gap between ethics of care and the sociology of welfare in the history of political philosophy, Yayo Okano traces the genealogy of ethics of care in the US women's liberation movement in the 1970s. The movement shed light on social relations that were subsumed in the liberalism that defined equality of individual responsibility as its fundamental principle (Okano 2015). Drawing from Carol Gilligan (1982), who established the theory of ethics of care through her research interviews with women struggling with abortion, Okano persistently deploys the term *recipient and provider* of care. Okano is careful not to use *tojisha* in her discussion. What distinguishes Okano from Ueno is that for Okano, it is impossible to distinguish self from the other because a self and the other are already woven into the fabric of interrelatedness and interdependency. Okano critiques liberalism's dominant ideology, which subsumes inter-dependent social lives into the realm of individual responsibility. Citing Eva Kittay's theory of dependency (1999), which critiques theories of liberal tradition as having relied on independence as a norm of personhood, Okano argues that the liberal theories developed in relation to the capitalist market economy served to eliminate and devalue people in interdependent relationships while relying upon them for their social existence (Okano 2015).

Citizenship and personhood within liberalism, according to Wendy Brown (1995), produce, require, and at the same time disavow their feminized opposites [the person who requires care] when the liberal subject emerges as pervasively masculine, not only in its founding exclusion and stratifications but in its contemporary discursive life. Okano is critical of the term autonomy (*jiritsu*) because people are already in relationships with the others, and without the other, one cannot exist.

The *Jiritsu* Model of Island Societies

Jiritsu of islands is the model that RSSI hoped to establish in the recently completed project "Conceptualizing '*Tosho chiiki kagaku kenkyu*' (RSSI) Toward the Creation of a *jiritsu* (Autonomy) Model of Island Societies" in 2019, as well as in the previously cited "Director's Greetings." *Jiritsu* has two homophones: both consist of two kanji characters and share the same first kanji character 自 (self). The latter kanji character determines the meaning of each word. 立 in the first word literally means to stand, so 自立 means to stand by oneself. From there, the word is defined as independence, non-dependency. The other is spelled 自律. 律 means to govern and order, so 自律 is defined as self-governing, a self with order. In other words, *jiritsu* (自立) emphasizes physical and economic independence while *jiritsu* (自律) emphasizes the aspects of volition for self-determination. RSSI makes a conscious attempt to use the latter because its aim differs from an economic or political independence to be achieved by islands themselves. RSSI aims not for economic independence or liberation of islands from interdependency but for study of islands with an understanding of co-dependency.

The concept of *jiritsu* has been developed in the field of social welfare in relation to the disabled people's movement. In Japan, it has been developed as an important ethical value. Social welfare is also a field where a number of discussions on subjectivity, individuality, and self-determination have been held. Even in this field, the importance of *jiritsu* has been taken for granted or as naturally desired. Many discussions tended to conclude that *jiritsu* is necessary because it enables liberation and dignity (Ishikawa 2009; Seika). Such arguments assume that the subject is already equipped with the abilities to perform *jiritsu*. Responses to solving the problem are twofold. First, representatives or agents of a subject represent and assist the subject's *jiritsu* if the subject is unable to determine methods of *jiritsu* or if the subject does not have the ability due to their physical condition. Second, it is understood that *jiritsu* is not achieved by the efforts of the subject alone but with assistance from others. However, Ishikawa argues that achievement of *jiritsu* depends on the subject's choices provided by an external environment and by a society's recognition of a person as *shutai* (subject). Thus, *jiritsu* and *shutai* are mutually exclusive concepts in social welfare studies, and by extension, they build a critical interrelatedness between the key words. This is one (or *the*) approach taken by RSSI.

Shutai and *Shutaisei* in RSSI

The first RSSI project, "Creating a New Island Studies: The Ryukyu Archipelago as the Hub That Links Japan to East Asia and Oceania," was "a vigorous initiative to pursue the proposition of realizing the sustainable development of island regions" (Research Institute for Islands and Sustainability n.d.).[3] The mission statement links Okinawa, Asia, and the islands of Oceania since "their relationship with the larger nation-states' military and foreign affairs in particular influenced the way they are today." The "new" (that RIIS promises) island studies initiative begins with "an examination of such positioning of the island regions not through the perspectives of the larger nation-states but by placing an emphasis on *shutaisei* of the island regions." This echoes how Baldacchino emphasizes that islands should be studied on their own terms. Such a mode of understanding encourages shifting the gaze of large nation-states away from small islands to a more nuanced exploration of epistemologies and ontologies of nation-states.

The director of RIIS, Yoko Fujita, states that "we hope to establish a foundation of theories and methods that place an emphasis on relationalities of island communities foregrounded on the subjectivity of the islanders as residents and researchers." Fujita labels this as the framework for the new island studies at RSSI (Fujita 2018a).

---

[3]This project consisted of four approaches: studies of Ryukyu-Okinawa; disciplinary studies of environment, culture, and society; transdisciplinary studies; and foundation of research information.

The term *subject* entered Japan through the introduction of European scholarship. That is to say, translated scholarship. However, it is not a simple act but rather the original meaning of a subject that is made hybrid. Sakai defines *shutai* as the body of enunciation, and it is "irreducible to the subject" (Morris 1997, p. xiii). As Morris interprets Sakai, "to address to different audiences is "necessarily to care about context, to respond to contingency, to admit limitation, to be willing repeatedly to differ" (Morris 1997, p. xxi). Translation of *subject* and *subjectivity* play a central role among contemporary Japanese critical scholars. In Japan, according to Sakai, subjectivity is translated into two determinations, "the epistemic subject (*shukan*) and the practical subject or agent of practice (*shutai*), in the context of the analysis of culture and cultural differences, although whether *shutai* can be subsumed under the generality of subjectivity is highly problematic and far from having been settled" (Sakai 1997, p. 119). Sakai (1997) argues that the Japanese word *shutai* is not neatly contained in the English translation of *subject*. Similarly, the Japanese word *shutaisei* includes independence and individuality. Moreover, Sakai and Miyoshi both argue that the notion of subjectivity was developed in the culturally specific context of the West while it was associated with war responsibility in post-Second World War II Japan. Shiho Satsuka also asserts that the issue of subjectivity "continues to be the key predicament for articulating a culturally meaningful model of 'citizen' in contemporary Japan" (2009, p.69). She argues in her analysis of populists' negotiation of Japanese subjectivity that the predicament of articulating Japanese subjectivity is reflective of the paradoxical position under the legacy of Cold War geopolitics in Asia and is conditioned by the US-Japan political, economic and military coalition (Satsuka 2009, p. 69). Satsuka delineates such a reflective position of Japanese subjectivity as constructed by the Japanese desire to see the West as a normative interlocutor.

Satsuka discusses the idea of the West as advanced, liberal, and dynamic, whereas Japan is seen as traditional, and argues that this idea propelled the pursuit of populist subjectivity in Japan but was not successfully integrated into the national discourse. For her, "in post-Cold War, post-bubble economy Japan, the pursuit of individual subjectivity and freedom was taken over by the new-liberal camp, with the discourse of self-responsibility and independence from state welfare and protection" (Satsuka 2009, p. 79).

Conclusion: Performativity and Islandness

As someone who has applied and considered gender theories in my own study of Okinawa, I cannot help but see the interrelated discourses between islands and gender. They both are considered the *other*, the former to the continent and the latter to men, respectively. Both identities are determined by dichotomies and hierarchical relationalities. Moreover, other studies on the Pacific Islands have also brought to light how islands have been historically feminized in the discourse of colonization (Shigematsu and Camacho 2010; Camacho 2019; Ginoza 2019). For instance, Camacho historicizes the Pacific Islands as treated like "an open frontier to be crossed, domesticated, occupied, and settled," and "often stereotypically gendered as feminine and vulnerable, thus requiring the protection of a masculine, military presence," therefore justifying an exploitation of indigenous peoples of the Pacific (Camacho 2019, p.xxxii). Judith Butler's genealogy of women who have been marginalized opens up the possibility for deconstructing the essentialist, fixated, and liberal position of identity of islandness. In *Gender Trouble* (2006), Butler defines gender not as a noun but as a verb performatively produced and compelled by the regulatory practices of gender coherence. Citing Simone de Beauvoir's famous line "one is not born, but, rather, becomes a woman," Butler argues the category of woman is constituted in acts; therefore, gender is not "a stable identity or locus of agency from which various acts proceeded; rather it is an identity tenuously constituted in time—an identity instituted through a stylized repetitions of acts" (1988, p. 519). Butler continues:

> gender is instituted through the stylization of the body and, hence, must be understood as the mundane way in which bodily gestures, movements, and enactments of various kinds constitute the illusion of an abiding gendered self. This formulation moves the conception of gender off the ground of a substantial model of identity to one that requires a conception of a constituted social temporality. Significantly, if gender is instituted through acts which are internally discontinuous, then the appearance of substance is precisely that, a constructed identity, a performative accomplishment which the mundane social audience, including the actors themselves, come to believe and to perform in the mode of belief. If the ground of gender identity is the stylized repetition of acts through time, and not a seemingly seamless identity, then the possibilities of gender transformation are to be found in the arbitrary relation between such acts, in the possibility of a different sort of repeating, in the breaking or subversive repetition of that style. (p. 519)

It is not only plausible to think about islandness with the theory of gender performativity, it is also interdisciplinarily engaging. If we consider identities endowed in the island and islander as processes, the island subject is not reduced or fixated to a submissive position to power. This shows that identity and islandness are not *a priori* categories but are open to the process of redefinition and rearticulating. This process requires critical engagement with the identity of islandness and islanders as aware and engaging of their identity's problems, contradictions, paradoxes, pain, and tensions. Islandness is unstable. To practice RSSI is to constantly surface that instability and fluidity.

To establish a new field that encompasses different regions, countries, and locations, a rich understanding and imagination that predicts limitations and dangers of transporting and applying theories or criticisms in different locations is important. Without such attention to the issue of reapplication, or without historicizing the translated criticism and accepting it without criticism or reasoned consideration, we may produce insensitivity and an avoidance of historical responsibility in place of a new order of knowledge and subjectivity, as well as one of compliance. Rather than grounding RSSI in uncritical acceptance and the framework of identification with the continent/larger nation-states, RSSI may produce a new presentation of islandness in order to overcome and move beyond the normal dichotomous modes islands have often been (re)presented in.

In this article I have delineated the keywords—*shutai, jiritsu,* and *tojisha.* They play an important role in articulating RIIS's model of Reginal Science for Small Islands. In the fields of humanities and social sciences, these words are discussed, theoretically as well as genealogically, in a mutually exclusive relation to each other. Genealogically, they resonate in the way they have been developed with liberalism's discourses of independence and self-reliance that constitute the core of dominant Western ideology. I have underscored the problematics of each word and suggested instead an understanding of co-dependency and co-construction of island models in order to disarticulate the internalized identities of islands as marginal, peripheral, and disadvantaged. This can be done well by applying a mode of heterolingual address to imagine a nonaggregate community of islands. In past approaches, islands are drawn or conceptualized as being female or feminine to a continent's maleness. Drawing on Butler's performativity of gender as a non-fixed identity, but as constituted in a social temporality, I have argued that island identity, or islandness, may also be reimagined through the instability and fluidity of islanders' performativity.

References

Baldacchino. (2008). Studying Islands: on whose terms? Some epistemological and methodological challenges to the pursuit of Island studies. *Island Studies Journal, 3*(1), 37–56.
Bhabha, H. (1995). *Location of culture.* London: Routledge.
Brown. W. (1993). Wounded attachments. *Political Theory, 21*(3), 390–410.
Brown, W. (1995). *States of injury: power and freedom in late modernity.* Princeton: Princeton University Press.
Butler, J. (1988). Performative acts and gender constitution: an essay in phenomenology and feminist theory. *Theatre Journal, 40*(4), 519–531.
Butler, J. (2006). *Gender trouble: feminism and the subversion of identity.* New York: Routledge.
Camacho, K.L. (2019). Transoceanic flows: pacific islander interventions across the american empire. *Amerasia Journal, 37,* iv–xxxiv.
Campbell, F.K. (2009). *Contours of Ableism.* Palgrave Macmillan.
Erevelles, N. (2011). *Disability and difference in global contexts: enabling a transformative body politic.* London: Palgrave Macmillan.
Fujita, Y. (2018a). From the director: regional science for small islands from Okinawa. *Research Institute for Islands and Sustainability.* https://riis.skr.u-ryukyu.ac.jp/vision.
Fujita, Y. (2018b). Purpose and philosophy. *Research Institute for Islands and Sustainability.* https://riis.skr.u-ryukyu.ac.jp/vision.
Gilligan, C. (1982). *In a different voice: psychological theory and women's development.* Cambridge: Harvard University Press.
Ginoza, A. (2019). Ajia pashifiku shiata to datsugunjishugi no bunka wo sozosuru gujishugi wo yurusanai josei netowaaku. Ikegami et.al. (Eds.). *Tosho chiiki kagaku toiu chosen.* Naha: Border Ink.
Ikegami, D., Sugimura, Y., Fujita, Y., & Motomura, M. (Eds.). (2019). *Tosho chiki kagaku toiu chosen.* Naha: Border Ink.
Ikegami, D., & Fujita, Y. (Eds). (2019). Introduction: 'Tosho chiiki-kagaku kenkyu' from Okinawa. In *Tosho chiiki-kagaku toiu Chosen.* Naha: Border ink.
Ishikawa, T. (2009). Noryoku to shiteno jiritsu: shakai fukushi ni okeru jiritsugainen to sono sonchou no kento. *Shakifukushigaku, 50*(2), 5–7.
Kittay, E. (1999). *Love's labor: essays on women, equality, and dependency.* New York: Routledge.

Morris, M. (1997). "Forward" to Sakai, N. *Translation and subjectivity.* University of Minnesota Press, Minneapolis.
Nakanishi, S., & Ueno, C. (2003). *Tojisha Shuken.* Iwanami Shoten, Chiyodaku.
Okano, Y. (2015). *Kea no rinri to fukushishakaigaku no kakyo ni musette: kea no rinri no sonzairon to shakairon yori* (Bridging the Gap between the Ethics of Care and Sociology of Welfare: Through Ontology and Social Theory of Care). *Fukushi Shakaigaku, 12,* 39–54.
Research Institute for Islands and Sustainability. (n.d.). Previous research philosophy. *Research Institute for Islands and Sustainability.*
Sakai, N. (1997). *Translation & subjectivity: On 'Japan' and cultural nationalism.* Minneapolis: University of Minnesota Press.
Satsuka, S. (2009). Populist cosmopolitanism: the predicament of subjectivity and the Japanese facination with overseas. *Inter-Asia Cultural Studies, 10* (1), 67–82.
Shigematsu, S., & Camacho, K.L. (Eds.). (2010). *Militarized currents: toward a decolonial future in Asia and the Pacific.* Minneapolis: University of Minnesota Press.
Ueno, C. (2010). Jenda kenkyu, tojishagaku no tachibakara (Gender studies from the standpoint of tojisha studies. *Shakaifukushigaku (The Japanese Society for the Study of Social Welfare), 51*(3), 132–135.
Ueno, C. (2011). *Kea no shakaigaku: tojishashuken no fukushi shakai e* (*Sociology of Care: Toward a Society of tojisha autonomy*). Shinjukuku: Ota Shupan.
Watanabe, T., Skrzypczak, E., & Snowden, P. (2017). *New Japanese english dictionary,* 5th ed. Chiyodaku: Kenkyusha.

Prospects for Island Studies

Islands as Safe Havens: Thinking About Security and Safety on Guåhan/Guam

Ronni Alexander

As a child, I remember being entranced by the watery worlds and island paradises described by such authors as James Michener and Robert Louis Stevenson. Later, I re-encountered them in the works of such artists as Paul Gauguin and in stories of World War Two like *Tales of the South Pacific* or the Japanese manga *Boken Dankichi* (Dankichi, an adventurous boy). The effect of these stories on generations of readers has been to create an image of the Pacific and its islands as a kind of "last paradise" that is not only romanticized and sexualized, but also militarized. Moreover, they have helped to create the understanding of small islands as unimportant, isolated, unable to manage their own affairs and expendable. This feminized othering has been used to legitimize, normalize and perpetuate colonial practices, including the use of islands for military bases and war fighting (Davis 2015). As a result, islands continue to play an important role as sites of military and colonial conquest.[1]

Like militarization, colonization is a gendered process that involves the denigration of a chaotic, barbaric, feminized other by a civilized, rational, educated and armed colonizer. Military power and sophistication are given as visible proof of colonial "superiority"; military bases with their modern technology become living symbols of difference, domination and desire. As minds and bodies become colonized, so do understandings of what it means to be and feel safe. This synergy between forces

---

[1] According to Baldachino (2004), the majority of non-self-governing territories are islands that seem to be relatively happy with their less-than-independent status. At the same time, local island communities often oppose hosting military bases (Lutz 2009). Guam is one of the seventeen entities that comprise the United Nations list of Non-Self-Governing Territories. A political status plebiscite offering three choices—independence, free association with the United States or statehood—has been called for, but it is unclear when or if the plebiscite will take place and, if it does, who will be allowed to participate in it. The elimination of U.S. bases from Guam is not a condition or stated goal for any of the groups established to promote the three options.

R. Alexander (✉)
Graduate School of International Cooperation Studies, Kobe University, Kobe, Japan
e-mail: alexroni@kobe-u.ac.jp

of colonization and of militarization is particularly strong in colonies such as Guam that were established primarily for strategic, rather than economic, reasons.

Taking a feminist perspective, this paper understands security to include both being and feeling safe, and interrogates the ways continuing colonization and militarization reinforce masculinities based on power, strength and the desire for protection through military means. Post-colonial and neo-colonial societies have been oppressed, their cultures, lands and languages desecrated and taken away. They have suffered attacks on their pride and joy in themselves and their culture, and over time the ways of the colonizer have become incorporated into their own cultures and behaviors. I suggest that this process includes the understanding of what it means to be safe and how that might be achieved. In other words, the colonization of bodies and minds includes understanding safety/protection within the worldview of their colonizers, even as they struggle to recreate the worldview that has been taken from them.

The following is primarily based on conversations on Guåhan/Guam[2] that took place between 2010 and 2018, and student responses to questionnaires about feelings of safety conducted in 2015 and 2017.[3] The questionnaire is used not to portray a single, objective story but rather to depict multiple voices and feelings, many of which often go unrecognized or are silenced.

Safety and Security on Guam

With about 30% of its land area under the control of U.S. military,[4] Guam is an interesting site for thinking about issues of militarization, safety and security. The presence of the bases puts Guam on the frontline of U.S. military activities in the Asia-Pacific region while at the same time the island is the homefront for the many island soldiers serving in the U.S. military at home and overseas (see Cohler 2017; Frain 2017). The military is everywhere on the island and the current military build-up is making its presence even more obvious. Soldiers in uniform are a common sight in tourist resorts and shopping centers, local television has military channels and local stores carry a variety of 'military-style' goods. The airport features a row of photographs of soldiers who have been killed in action in U.S. wars, while the bookstore carries counting books for each of the branches of the military, enabling children to learn their numbers by counting guns and tanks. The current military expansion

---

[2]Guåhan is the indigenous name for the U.S. colony of Guam. Here I will use the term 'Guam' to emphasize the colonized status. The indigenous people are Chamoru (alternative spellings: CHamoru, Chamorro); (Taitano 2014).

[3]The 2015 questionnaire included responses from Okinawa, mainland Japan (Kansai) and the United States. Here only responses from Guam are used.

[4]Most of this land was taken without fair remuneration from local owners after WWII or acquired through eminent domain.

means that expensive military housing complexes are replacing local homes and neighborhoods, and that the bases are more visible than usual.[5]

U.S. military colonization of Guam began in 1898 at the end of the Spanish American War and with the exception of three years of Japanese occupation during World War II, has continued until today. After the War, governance of Guam was returned to the U.S. Navy.

In 1950, the Organic Act of Guam took jurisdiction of Guam from the Navy to the Department of the Interior and gave residents U.S. citizenship, but not the right to vote for president or to have voting representation in Congress. The bases remained, growing and shrinking according to U.S. military objectives. Today the military is one of the two main industries, the other being tourism. Most families have a history of military service, adding to the sense that the military is an obvious career choice. According to one veteran, "Joining the military is a rite of passage, a way to prove yourself as an adult. But the reality is that people from Guam are looked down on in the military because they are from a colony. They have to defend themselves and do better than anyone else…" (BB, 2018.5).

Even today, many residents remain grateful and feel indebted to the U.S. for "saving" Guam from the Japanese at the end of the war and the legacy of "liberation" is still important. (Diaz 2001; Perez 2002; Bevacqua 2010). For example, one woman explained that while she knew the bases are a threat, "In 1941 the Japanese came because they knew the U.S. wasn't there to protect us" (C. conversation, 2016.5). Even many high school students explain that without the bases, Guam would be occupied by North Korea or China or perhaps some other power (Southern High School, 2017.9). A conversation between two adult friends is representative of the debate. One told said that it was safer because of the presence of the military. She said they would help if there were a disaster or an attack, asking "How can we protect ourselves if North Korea or China drop bombs on us?" (A. conversation, 2016.5). Her friend, B, took the opposite position. "The military cares about their own safety, not ours" (B. conversation, 2016.5).

Whose Security? Whose Safety?

According to the United States, Guam is the "tip of the spear" of American power projection in the Asia- Pacific region. Because Guam is an American territory, it is different from most other sites in the network of U.S. bases; there is no need for Status of Forces agreements (SOFA) or negotiations with a foreign government. As people do not have the right to vote for president or a voting representative in Congress, the U.S. government does not really need to worry about what the people of Guam

---

[5] An extensive military build-up that includes relocation of 3000 U.S. Marines and their dependents from Okinawa to Guam is currently underway on Guam. The flagrant disregard of cultural and environmental issues has caused some resentment, but many feel safer with the bases there (Alexander 2013, 2016a, b; Na'Puti and Bevacqua 2015; Nagashima 2015, 2018).

think. In the words of one Pacific Air Forces commander, "Guam, first of all, is U.S. territory…. I don't need overflight rights. I don't need landing rights. I always have permission to go to Guam. It might as well be California or New Jersey" (Brooke 2004).

Military bases contribute to militarization, an on-going process that brings the military into a range of social relations otherwise unrelated to war and/or war-making, making them seem 'natural' or 'normal'. Real and imagined symbols of military strength like military bases depend on articulations of insecurity to which violent responses are believed to be the most effective response.[6] Bases help to construct and reinforce securitized understandings of what is and is not safe and to normalize military solutions, making them appear to be the only and/or most reasonable choice. This militarization of understandings of safety is not limited to soldiers, but also affects people who work at military bases or live in the surrounding communities (Vine 2015).

Security is a difficult concept that requires asking who is protecting whom from what, and is linked with socially constructed understandings of vulnerability and protection that generally involve identity. In colonies, particularly those established for primarily military reasons, just as various bodily and social practices are colonized, so are understandings of security and safety. Notions of identity, equality and difference are particularly powerful tools for militarization and are used to motivate soldiers to fight and communities to support them. Gendered understandings of safety and of war as necessary for protection of the homeland, family and perhaps especially women and children are particularly potent symbols.

Feminist scholarship on military bases (Lutz 2009; Höhn and Moon 2010; Sturdevant and Stoltzfus 1992), aspects of peace and war (Hunt and Rygiel 2006; Sjorberg and Gentry 2007; McSorley 2013; Sylvester 2002, 2013) and work from Pacific scholars stressing oceanic connections among islands (Hau'ofa 1994; Teiwa and Slatter 2013) have problematized binaries and pointed out the gendered aspects of colonization, militarization and security, as well as the importance of multiple voices and stories. Duffield (2001), Dillon and Julian (2009), Burke (2002) and others have shown how our understanding of what it is to be secure is bound to what it is to be insecure. This also includes our understandings of gender and how, when and with whom we feel safe.

In traditional Western understandings of the nation-state, the honor of being a soldier and sacrificing one's life for the protection of the nation was limited to men of a particular race and class. Today many countries have women serving in the military and some countries, the United States included, now allow women to serve on the front line in many capacities that were formerly indisputably reserved for men. This has not necessarily changed the gendered binaries on which the military is based; rather it means that women must also conform to, or at least pass tests of, military masculinity. In the words of one veteran from Guam, "If she can hold her weight as a man, I have no objection to having a woman serving with me" (BJ

---

[6]The recent North Korean missile tests are a good example; Guam was said to be safe because of the bases, but if there had not been any bases, there would not have been any threat.

interview, 2018.5). This supports Butler's (2004) assertion that gender is a norm that "operates within social practices as the implicit standard of *normalization*" (italics in the original, p. 41).

Emotions are important for thinking about safety. They are an integral part of identity, and like identities they may be multiple, conflicting and inconsistent. Most work on security conflate being and feeling safe,[7] but it is possible to be safe and not feel that way, and to feel safe without really being so. Scholars and policy makers use fear and feelings of insecurity as justification for militarization and many acts of aggression, but this is generally not acknowledged as the manipulation of emotion. Rather, emphasis is given to the importance of rationality while emotion is treated as an obstacle to good decision making. In fact, emotions such as fear are very much a part of the way people view and interpret the world and are an important and increasingly acknowledged part of identity, culture and politics (see for example Lutz and Abu-Lughod 2008; Crawford 2000, 2014; Hutchison and Bleiker 2008, 2014; Ahmed 2014; Mercer 2014; McDermott 2014).

Some of the most important and perhaps most manipulated emotions related to place, space and identity are those related to fear. But in the field of international relations, the link to emotion is so normalized that it can be hard to see. Emotion becomes part of a masculine/feminine dichotomy in which masculinized security involving rational, objective thinking and assertive action is said to keep us safe, while irrational, impulsive, or feminine ways of knowing are dangerous. Regardless of gender, in the institutionalized willingness to sacrifice their lives for the protection of others, soldiers become important symbols of military understandings of security and safety rooted in rationality, physical strength and cool toughness.

In the masculinized imaginary of the military, there is little room for emotion. As Enloe (2000, 2007) has shown, militaries rely on othering of the feminine for the construction and maintenance military masculinities. The hyper-rational and hyper-masculine states "denigrate anything smacking of the feminine, including a sense of welfare and compassion for all, natives and aliens alike" (Ling 2014, 580). While some scholars have begun to identify multiple masculinities in the military (Higate 2003; Whitworth 2005; Gonzalez 2010; Eichler 2012) and others identify violent femininities (Sjorberg and Gentry 2007), militaries continue to represent particular relations of masculinity and power.

Talking About Being and Feeling Safe

In 2015, I asked colleagues on Guam to distribute a simple questionnaire to 65 of their students. In September 2017, I used the same questionnaire with 58 university and high school students. At that time, North Korea was conducting nuclear-capable

---

[7]The language of being and feeling safe is difficult. While often "safety" is used to refer to personal safety as opposed to national "security", here "being safe" refers to the state of physical safety/security while "feeling safe" refers to the emotion of feeling safe/secure.

missile tests and claiming to be targeting Guam. The purpose of the questionnaire was not to paint a complete picture and/or generalize about how young people on Guam feel about the U.S. bases, but rather to show how particular people in particular situations (e.g. University of Guam students taking Chamoru language classes at a particular moment in time) expressed their feelings and concerns. The following is a brief summary of their responses.

Characterizing Guam

Respondents were first asked to choose from a list of words those that most closely describe their image of Guam.[8] By far the most frequently chosen option was "A place I love," followed by "A place I want to protect." This is true for both the original questionnaire (2015) and the second (2017), regardless of the respondent's attitude toward the bases and security. When asked why, many from both groups emphasized their love for Guam as their home and its importance for Chamoru culture. For example, "Because it is my birth place and my culture is dying within the American presence" (G2-01, 2016) or "This is my home. The Chamorro island is unique and if this island was ever to vanish, there wouldn't be anything left like us" (UGO9_2018). Fewer numbers in both groups suggested it was a peaceful place, military base and center of Chamorro culture. A few students in the 2015 group, and fewer in the 2017 group, characterized it as a place they wanted to leave, commenting that they wanted to see the world and gain experience. A very small number in both groups said it was a dangerous place. In 2015, reasons given were environmental threats from the bases, while in 2017 respondents noted military threats from North Korea and/or China (2018: SHP4-4; UOG_14; SHP4). In the context of this paper, these characterizations speak to the strong sense of place of these young people and its importance in their identities.

Do the Bases Make You Safe?

In 2015, most students on Guam thought the U.S. bases both made them safe and made them feel safe. "They protect us from any invaders" (G1-16); "They are to house military personnel which would in turn provide safety from terrorists/harmful military groups" (G2-10); "because it (Guam) is positioned strategically" (G1-09). At the same time, some thought the bases made them unsafe. For example, "Invitation for war" (G1-03); "Targeted by other countries" (G1-23) or "Because of the bases, it

---

[8]Questionnaire choices for ways to describe Guam: Island resort; military base; U.S. colony; Center of Chamorro culture; An island necessary for U.S. defense; U.S. possession; A place I love; A dangerous place; A place I want to leave; A place I want to protect; A peaceful place; Other

puts U.S. as a target to others. For instance, what North Korea threatened to do (e.g. attack Guam) was because Guam had missiles located on bases here" (G2-28).

In 2017, many students cited the strength of the U.S. military. For example, "US military is the strongest in the world; no other country can match its personnel strength and technological capability" (2017: UOG4). They also stressed the ability of the military to protect them from threats, as in, "We are protected from anything that may threaten us" (2017: SH4_20). A few students mention North Korea and/or China by name, but most referred to general "threats" or "attacks". What is interesting is that while most people in both groups said that the bases made them safe, in 2017 many also said that they weren't sure if the bases made them feel safe or said they did not *feel* safe. For example, the goal of the U.S. "isn't to protect Guam" (2017: UOG14).

While the bases make more respondents feel safe than unsafe, some in both groups expressed ambivalence: "They make me feel like we are protected = safe. But I also feel unsafe at the same time because they are targets" (2015: G1-01). Some expressed ambiguity: "I feel like even though we are targeted because of our bases, we are better prepared for attacks because of them" '2017': UOG5) or "I do feel safe but at the same time I feel like they are using us" (2017: SH16). Responses such as these underscore the contradictions represented by the bases: many students express that there is strength and safety in numbers and firepower. At the same time, they wonder if the U.S. is really interested in protecting them, or whether they are being used as a shield to protect the U.S. Many see the bases as a kind of insurance—it is better to have them just in case something bad happens—but few expressed what exactly the 'something bad' would look like.

In 2017, I also held a workshop with students at Southern High School where issues of being and feeling safe were discussed. While repeating over and over that the U.S. has the strongest military in the world so they are safe, the written notes from their group discussions revealed ambivalence. For example, "Nobody is taking the North Korean threat seriously or trying hard to put a stop to it" or "The North Korean 'threat' has affected us dramatically because due to the threat Japanese school exchanges cancelled out on us" (2017, Southern High School WS). One student told me that her family had a boat ready, just in case they needed to escape, and another said that she doesn't like to think about the threat because Guam does not have an evacuation plan and so everyone would probably die (2017.9 Southern High School).

Where Do You Feel Safe?

Several questions were about when/where people felt peace, and when/where they felt and/or were safe. Most respondents from both years, as well as most people with whom I conversed, said they felt most safe and peaceful at home or with their families. The second most frequent response was feeling safest on military bases. A typical response from 2015 was, "I feel peaceful and safe at home, church & on military installations on island (G1-25). Many people also said they felt peace (and also safe) when with friends. In 2017, the responses were similar, although a few people said that they do not ever feel safe. Some people said that being at the beach or in nature made them feel peaceful, and some said they felt safe in school.

Obligation to Contribute to Security

As citizenship, military service and gender are linked, several questions addressed whether and why respondents do or do not feel a responsibility to contribute to the security of the United States and, for those who do, what that contribution might entail. I was interested to see whether people on Guam felt that giving one's life in defense of the U.S. was a necessary part of being a citizen.

In 2015, quite a few people said they were currently, or had previously been in and/or affiliated with the military and that was how they fulfilled their responsibility. Others felt responsibility because they live in a U.S. territory (G2-29) or because "The U.S. protects Guam" (G2-25). Others felt that "I just don't really feel like it is/would be appreciated" (G1-24) or that "I don't feel like I'm even a part of America" (G1-26). One person captured the ambivalence of both being, and not being, American: "Again I grew up with a weird dichotomy of nationality. I am told the best way to contribute to safety and peace is to enlist in the military but I am also told to aspire toward the American dream of chasing my dreams" (G2-33).

In 2017, while a few people said they felt a responsibility, most either denied having that feeling, or indicated they felt powerless as citizens, or were questioning. For example, "Guam doesn't have much say in what goes on in the US, ex: we don't vote for the president" (2017: UOG19) or "in my opinion, the US, as the strongest nation in the world, should already have enough to sustain themselves" (2017: UOG 18). As to what they might do as responsible citizens, one student said, "I'm not sure; we don't really have a vote…probably just to join the military" (2017: SH4_18).

Conclusion: Gender, Militarization and Being/Feeling Safe on Guam

Guam is one of many feminized and militarized islands being used for protection of a distant homeland. This paper suggested colonization and militarization affects understandings of being and feeling safe, and looked at the implications for people on Guam. It has shown that continuing colonization and militarization is gendered and reinforces masculinities based on power, strength and the desire for protection through military means. At the same time, some people understand and articulate the contradictions inherent in having the military on Guam to protect them.

If Guam did not have bases, there would probably be no reason for North Korea or any other country to attack it and no need for defense against malicious terrorists or armies. Yet many people say they feel safer knowing the military is there, even though some also acknowledged that the U.S. military presence was also the cause of the problem. This collective emotion of feeling unsafe has somehow become entwined with an understanding of masculinized military power as ensuring peace and security, leading to the belief that people would feel even less safe without the bases. This is one way that overseas military bases reproduce a discourse of insecurity which allows for their continued, if contested, existence.

Another aspect of this is the reproduction of U.S. military understandings of safety. Not only do people stress the strategic importance of Guam, but they also show approval for the strength of the U.S. military. In 2017, when explaining why they felt the U.S. military presence made them safe and/or feel safe, nobody made any reference to, or expressed concern about, what might happen if the U.S. did in fact shoot down North Korean missiles and Japan or other islands happened to get caught in the cross fire. In other words, people on Guam were reiterating American understandings of security, rather than expressing alternative ways of creating/protecting themselves

This chapter has demonstrated that for many on Guam, being safe and feeling safe might in the end only be possible in unsafe ways—through military protection—a reflection of not only the ways people view peace and security but also of how their continued support of 'protection' by militarily powerful 'others' is maintained. In order for people on Guam to imagine security that is not based on insecurity, decolonization is essential, but for that, they need to be able to re-imagine their island as Guåhan rather than as Guam, and to envision it without military bases. This means looking within, but also imagining their interconnections with people in places outside those borders.

Colonization and militarization on Guam have produced an internalization of security to mean "safety-as-protection" and pervasive normalization of military solutions as natural and successful. The search for ways to truly be and feel safe must begin with an endogenous process of de-colonization and de-militarization that comes from individual and collective bodies and includes a re-creation of gendered relations. That search must challenge meta-narratives of security that deny the importance of feelings of safety and allow for the development of ways to establish living places and spaces where people can feel and be safe.

References

Ahmed, S. (2014). *The cultural politics of emotion* (2nd ed.). Edinburgh University Press.
Alexander, R. (2013). Militarization and Identity on Guam: Exploring intersections of indigeneity, gender and security. *Journal of International Cooperation Studies, 21*(1), 1–22.
Alexander, R. (2016a). Living with the Fence: militarization and military spaces on Guahan/Guam. *Gender, Place and Culture, 23*(6), 869–882. First published on-line in 2015.
Alexander, R. (2016b). Peace in the Pacific? Living strategic colonialism in the American Pacific. Unpublished paper presented to the ISA Annual Conference, March 2016, Atlanta, GA.
Baldacchino, G. (2004). The coming of age of island studies. *Tijdschrift voor Economische en Sociale Geografie, 95*(3), 272–283.

Bevacqua, M.L. (2010). The exceptional life and death of a chamorro soldier: tracing the militarization of desire in Guam, U.S.A. In S. Shigematsu & C. Keith (Eds.), *Militarized Currents: toward a decolonized future in Asia and the Pacific* (pp. 33–62). Minneapolis: University of Minnesota Press.

Brooke, J. (2004). Looking for friendly overseas base, pentagon finds it already has one. *New York Times*. 7 April.

Burke, A. (2002). Aporias of security. *Alternatives: Global, Local, Political*, 27(1), 1–27.

Butler, J. (2004). *Undoing Gender*. New York and London: Routledge.

Cohler, D. (2017). Introduction: homefront frontlines and transnational geometries of empire and resistance. *Feminist Formations*, 29(1) (Spring), vii–xvii.

Crawford, N.C. (2000). The passion of world politics: propositions on emotion and emotional relationships. *International Security*, 24(4), 116–156. Posted Online March 29, 2006.

Crawford, N.C. (2014). Institutionalizing passion in world politics: fear and empathy. *International Theory*, 6(3), 535–557. Published online: 09 October 2014.

Davis, S. (2015). *The empires' edge: militarization, resistance, and transcending hegemony in the Pacific*. Athens: The University of Georgia Press.

Diaz, V.M. (2001). Deliberating 'Liberation Day': identity, history, memory, and war in Guam. In T. Fujitani, M. Geoffrey, M. White, Y. Lisa (Eds.), *Perilous memories: the Asia-Pacific war* (pp. 155–180). Durham: Duke University Press.

Dillon, M., & Julian, R. (2009). *The liberal way of war: killing to make life live*. Routledge.

Duffield, M. (2001). *Global governance and the new wars: the merging of development and security*. London: Zed Press.

Eichler, M. (2012). *Militarizing men: gender, conscription, and war in post-soviet Russia*. Stanford: Stanford University Press.

Enloe, C. (2000). *Manuevers: the international politics of militarizing women's lives*. Berkeley: University of California Press.

Enloe, C. (2007). *Globalization and militarism: feminists make the link*. Rowman and Littlefield.

Frain, S. C. (2017). Women's resistance in the marianas archipelago: a US colonial homefront and militarized frontline. *Feminist Formations*, 29(1), 97–135.

Gonzalez, V. V. (2010). Touring military masculinities: U.S.-Philippines circuits of sacrifice and gratitude in Corregidor and Bataan. In S. Shigematsu & K. Camacho (Eds.), *Militarized currents: toward a decolonized future in Asia and the Pacific* (pp. 63–90). Minneapolis: University of Minnesota Press.

Hau'ofa, E. (1994). Our sea of islands. *The Contemporary Pacific*, 6(1), 148–61.

Higate, P. (2003). *Military masculinities: identity and the state*. Westport, CT: Praeger.

Höhn, M., & Moon, S. (Eds.). (2010). *Over there, living with the U.S. military from world war two to the present*. Durham: Duke University Press.

Hunt, K., & Kim, R. (Eds.). (2006). *(En)Gendering the war on terror*. Ashgate.

Hutchison, E., & Blei, R. (2014). Theorizing emotions in world politics. *International Theory*, 6, 491–514.

Hutchison, E., & Bleiker, R. (2008). Fear no more: emotions and world politics. *Review of International Studies*, 34, 115–135.

Ling, L.H.M. (2014). Decolonizing the international: towards multiple emotional worlds. Forum: emotions and world politics. *International Theory*, 6(3), 579–583. Retrieved May 20, 2017, from https://www.cambridge.org/core/services/aop-cambridge-core/content/view/026BC7C8D8B25151DB4862D3E6F08DFA/S175297191400030Xa.pdf/decolonizing-the-international-towards-multiple-emotional-worlds.pdf.

Lutz, C. (Ed.). (2009). *The bases of empire: the global struggle against U.S. military posts*. Pluto Press.

Lutz, C. A., & Abu-Lighod, L. (2008). *Language and the politics of emotion*. Cambridge: Cambridge University Press.

McDermott, R. (2014). The body doesn't lie: a somatic approach to the study of emotions in world politics. *International Theory*, 6, 557–562.

McSorley, K. (Ed.). (2013). *War and the body: militarisation, practice and experience.* Routledge.
Mercer, J. (2014). Feeling like a state: social emotion and identity. *International Theory, 6,* 515–535.
Na'Puti, T., & Bevacqua, M.L. (2015). Militarization and resistance from Guahan; defending Pagat. *American Quarterly, 67*(3), pp. 837–858.
Nagashima, R. (長島怜央) (2015). 『アメリカとグアム―植民地主義、レイシズム、先住民』有信堂.
Nagashima, R. (長島怜央). (2018). 「標的のアメリカ植民地―北朝鮮の核・ミサイル問題におけるグアムと北マリアナ諸島の人々」アジア・アフリカ研究 第2号(通巻428号), pp. 31–56.
Perez, M. (2002). Pacific identities beyond U.S. racial formations. the case of Chamorro ambivalence and flux. *Social Identities, 8*(3), 457–479.
Sjoberg, L., & Gentry, C. (2007). *Mothers, monsters, whores: women's violence in global politics.* Monsters, Whores: Zed Books.
Sturdevant, S.P., & Stoltzfus, B. (1992). *Let the good times roll: prostitution and the U.S. military in Asia.* New York: The New Press.
Sylvester, C. (2002). *Feminist international relations: an unfinished journey.* Cambridge University Press.
Sylvester, C. (2013). Experiencing war: a challenge for international relations. *Cambridge Review of International Affairs, 26*(4), 669–674.
Taitano, G.E. (2014). Chamorro vs. Chamoru. *Guampedia.* Retrieved May 20, 2017, from http://www.guampedia.com/chamorro-vs-chamoru/.
Teiwa, T., & Slatter, C. (2013). *Samting nating:* pacific waves at the margins of feminist security studies. *International Studies Perspectives, 14,* 447–450.
Vine, D. (2015). *Base nation: how U.S. military bases abroad harm America and the world.* New York: Metropolitan Books.
Whitworth, S. (2005). Militarized Masculinities and the Politics of Peacekeeping: The Canadian Case. In K. Booth (Ed.), *Critical security studies in world politics* (pp. 89–106). Boulder, CO: Lynne Rienner Publishers.

Island Studies and the US Militarism of the Pacific

Elizabeth DeLoughrey

Last summer (2018), the largest maritime exercise in history took place in the Pacific Ocean. Twenty-five thousand military personnel descended on the ocean area between the Hawaiian archipelago and southern California to participate in "war games," including nearly fifty naval ships, two-hundred aircraft, and five submarines. The 26th biennial Rim of the Pacific (RIMPAC) exercise was comprised of the military forces of twenty-five predominantly Pacific Rim nations, with the notable exceptions of China and Russia.[1] The theme of the five-week long RIMPAC 2018 was "Capable, Adaptive, Partners;" its purpose, according to the US Navy, is to "demonstrate the inherent flexibility of maritime forces" in regards to everything from disaster relief to "sea control and complex warfighting."[2] Past years have included exercises like sinking warships; this year's agenda lists amphibious operations, explosive ordinance disposal, mine clearance, diving and salvage work, as well as the live firing of anti-ship and naval-strike missiles.[3] While US imperial interests in the region have categorized the largest ocean on our planet as an "American Lake," military incursion by the PRC into the Spratly Islands has increased the Pentagon's concern that the Pacific is rapidly becoming a "Chinese Lake" and incentivizing military build-up in the region.[4]

Scholarship in island studies has helped bring forward the relationship between land and sea, allowing us to examine the ways in which oceans have been territorialized by the US military, which coined the term "island hopping." Yet most work in island studies has focused on the land rather than the seas. I seek to place island studies in a conversation with scholars calling for a "critical ocean studies" for the 21st century who have fathomed the oceanic depths in relationship to feminist and Indigenous epistemologies and multispecies studies.[5]

Decades of scholarship has positioned the ocean as *aqua nullius*; a blank space across which a diasporic masculinity might be forged.[6] More recent scholarship has

E. DeLoughrey (✉)
Department of English, University of California, Los Angeles, CA 90095, USA
e-mail: deloughrey@humnet.ucla.edu

animated the oceanic realm as matter and ontology rather than treating the sea as an inert backdrop.[7] In this oceanic turn, the concepts of fluidity, flow, routes, and mobility have been emphasized over other, less poetic terms such as blue water navies, mobile offshore bases (MOBs), high seas exclusion zones, sea lanes of communication (SLOCs) and maritime "choke points." I argue that this strategic military grammar is equally vital for a 21st century critical ocean studies for island studies and the Anthropocene. Perhaps because it does not lend itself to an easy poetics, the militarization of the seas is underrepresented in scholarship and literature emerging from what is increasingly called the blue or oceanic humanities. This paper builds on previous calls for "a critical militarisation studies" (CMS); one that weaves the complex histories of state violence in the region in relation to issues of ethnicity, indigeneity, gender and sexuality. This vital demilitarization work, including the important contributions about anti-base militarism in Okinawa by Ayano Ginoza, informs my engagement with CMS as well as critical ocean studies.

Island studies scholars have been at the forefront in bringing a critical discourse about militarism into the academy. There is ample, in fact, overwhelming visual documentation of the militarism of the oceans. The US Navy has long devoted their budgets to the visual reproduction of their military power at sea. This is evident in the spectacular photography and films of their nuclear weapons testing in the Pacific Islands between 1946 and 1962. Here is but one example taken from Operation Dominic, where the US launched 31 nuclear weapons in the Pacific Islands and their waters in the wake of the Bay of Pigs invasion. This particular image is of the 20-ton anti-submarine nuclear weapon named "Swordfish," fired in 1962 from the *USS Agerholm*.

The visual reproduction of the US military's destructive power over sea and airspace—the global commons—continues today in their social media blitz about exercises at RIMPAC including their twitter feed (see #ShipsofRIMPAC). When it comes to the military, it seems that hypervisibility produces invisibility. In this way the US military remains—to most of academia—hidden in plain sight.

Although marine biologists may point out that "every breath we take is linked to the sea" and that planet Earth is in fact "a *marine* habitat,"[8] another kind of planetary metabolism is equally constitutive—American militarization of the oceans is foundational to maintaining the global energy supply that undergirds what Andreas Malm and Jason Moore have termed the Capitalocene.[9] Over sixty percent of the world's oil supply is shipped by sea and over twenty percent of the Pentagon's budget goes to securing it.[10] Securing the flow of oil has been a vital American naval strategy—not to say mission—since the 1970s.[11] In fact, some have warned that there is a "dangerous feedback loop between war and global warming" because the Pentagon, in securing its energy interests through extensive maritime and overseas base networks—estimated at over 7,000—is the world's single largest consumer of energy, and the biggest institutional contributor to global carbon emissions.[12]

The US Navy and its associated Air Force emit some of the dirtiest bunker and jet fuels in order to secure the safe passage of maritime oil transportation; this energy in turn is consumed and emitted by the military in disproportionate rates to any nation.[13] This fuel cycle is common knowledge in military circles; in fact the Pentagon was exempted from all the major international climate accords *and* from domestic carbon emission legislation.[14] It should concern us that "militarism is the most oil exhaustive activity on the planet."[15]

Transoceanic militarism—via sail, coal, steam, or nuclear-powered ships and submarines, has long been tied to global energy sources, masculinity, and state power. Hosted by the US Navy's Pacific Fleet since 1971, RIMPAC's oceanic war games have been a way to make visible what the 19th-century naval historian Alfred Thayer Mahan famously termed "the influence of sea power upon history." While Captain Mahan recognized the sea as a commons, and even as "the common birthright of all people" (42), he spent his influential career advocating "the development of sea power" (43) for the United States which was critical to its 19th-century expansion into an "insular empire" spanning from Puerto Rico to the Philippines.[16] Mahan's political influence helped convince US leadership about the importance of sea and wind currents in positioning Hawai'i as a vital naval base and coal refueling station as well as a bulwark against China.[17] The 1898 annexations reflected the rise of American naval imperialism, where newly acquired colonies like Guam (Guahån) were administered by the US Navy as if the island were a ship. A few years later islands and atolls like American Samoa were claimed as essential to fuel the US military and ruled by the Navy as coaling stations.[18] From the US annexation of Micronesia in 1947, creating the Trust Territory of the Pacific, to the current US practice of claiming *permanent military exclusion zones* on the high seas in order to test weapons–nowhere has this sea power been more apparent than in the world's largest ocean.[19]

The Pacific Ocean as defined by geographers covers one third of the world's surface area (63 million square miles), but to the US military it extends all the way to the west coast of India, a nation that now participates in RIMPAC and represents the largest naval force in South Asia. Significantly, in spring 2018 the US military renamed its largest base, the Hawaiian-located Pacific Command, to the "US Indo-Pacific Command" (USINDOPACOM) in recognition of its new maritime regime, which has expanded to 100 million square miles, or a stunning "fifty-two percent of the Earth's surface."[20] This is an unprecedented naval territorialism that was almost entirely overlooked in the press and has not yet factored into any scholarly discussions of the Anthropocene or oceanic humanities.

Like the expansion into the Pacific Islands in the 19th century, the US Navy's inclusion of the Indian Ocean in their definition of the Pacific derives from strategies of energy security. There are five vital "sea lines of communication" (SLOCs) that connect both oceans through a lifeline of oil shipments from the Middle East. According to the US Navy website, "RIMPAC is a unique training opportunity that helps participants foster and sustain the cooperative relationships that are critical to ensuring the safety of sea lanes and security on the world's oceans."[21] Because the majority of oil exports are over water, US energy policy has become increasingly militarized and secured by the Navy, which is the largest oceanic force on the planet. Scholars such as Michael Klare have characterized the US military since the 2003 Iraq war "as a global oil protection service, guarding pipelines, refineries, and loading facilities in the Middle East and elsewhere."[22] US Naval spokespeople readily admit that RIMPAC is an exercise in "power projection," a political and military strategy to use the instruments of state power quickly and effectively in widely dispersed locations far from the territorial state. Others might use the term transoceanic empire, with the recognition that much of this (nuclear) power is also submarine. Fluidity, mobility, adaptability, and flux—all terms associated with neoliberal globalization regimes as well as the oceanic humanities—are also key terms and strategies of 21st century maritime militarism.

Postcolonial scholars recognize that Cold War politics reshaped academic funding channels, training and hiring, the formulation of departments (such as area studies), and even their vocabularies. Thus when the US annexed territories in Micronesia and put them in the hands of the Navy, academic funding was made available to anthropologists, including Margaret Mead, to study Pacific Islander cultures.[23] The rise of a 21st-century oceanic humanities would benefit from an interrogation of how it may participate in, mitigate, or challenge larger strategic interests, examining how our current geopolitics shape academic discourse, not to say funding. Simon Winchester, writing in the early 1990s at the inception of globalization studies, described what he called "Pacific Rising," noting that this oceanic turn—following the logic of transnational capital—was "quite simply" about "*power*" (27). And that power was represented, celebrated, and contested in the rise of globalization studies, Asia-Pacific studies, and Indigenous Pacific studies, fields largely informed by new models and knowledges of the sea.[24]

While globalization studies of the late 20th century emerged in relationship to the rise of transoceanic capital and its flows of "liquid modernity," to borrow from

Zygmunt Bauman, we might raise the question as to how 21st century articulations of an oceanic humanities and a turn to what some are calling "hydro-criticism" might be informed by larger geopolitical shifts.[25] Since the Obama era there has been a US "Pacific pivot" that includes transoceanic militarism as well as a trade treaty that, according to Robert Reich, entails "forty percent of the world economy."[26] The Trans-Pacific Partnership (TPP)—critiqued as "NAFTA on steroids"—includes an attempt to solidify transnational energy and sea-bed mining interests over state environmental protections.[27] Of course, its key security agents are naval forces, particularly evident in the highly contested military "mega buildup" on Guahån, one of the Navy's many "lily pads" and refueling stations that some American pilots refer to as "the world's largest gas station."[28] In a remarkable erasure of Indigenous presence, many militarized islands and atolls of American-occupied Micronesia have been referred to "unsinkable aircraft carriers" since the World War II era.[29] This is how militarized "ocean-space" is transformed into a "force-field," a term Philip Steinberg uses to describe when the "ideological value of sea power" merges with "the key role of a strong 'blue-water' fleet in troop mobility, naval warfare" in the quest towards the "domination of distant lands".[30]

Building on CMS, we might think about island studies in terms of hydro-power, defined as energy, force, militarism, and empire. As we turn to new sites of planetary expansion, flow, energy, and fluidity we might ask, from the perspective of Anthropocene studies, where is the body of literature and scholarship responding to these global shifts in hydro-power? Where is the literary, artistic, and cultural critique of an aquatic territorialism of fifty-two percent of the Earth's surface?

The novelist Amitav Ghosh raises similar questions in tracing out the relationship between energy, narrative, and the Anthropocene. In *The Great Derangement: Climate Change and the Unthinkable,* he builds upon his earlier observation that, given the ways in which the world economy is undergird by oil, it's peculiar that there have been so few "petrofiction" novels. Nearly twenty-five years later, he asks why, in an era of disastrous climate change, we see so few literary responses that take on its global, disastrous scope. I believe Ghosh's observations are relevant to calling attention to the gap in oceanic studies scholarship and literary production about US militarism more broadly. Ghosh concludes that the European novel narrowed the scale of "serious fiction" to an anthropocentric focus as well as a time scale that cannot account for deep time. Thus when faced with catastrophic climate change, or nonhuman agency, the European-derived novel has difficulty engaging the "uncanny intimacy of our relationship with the nonhuman". He raises a provocatively maritime question: "Are the currents of global warming too wild to be navigated in the accustomed barques of narration?"

Of course, no other region on the planet has been so deeply engaged with oceanic metaphors as Indigenous Pacific Islander studies, which has drawn extensively upon the image of the voyaging canoe as a vessel of the people and metaphor for navigating the challenges of globalization and ongoing colonialism.[31] Ghosh may had come to different conclusions if he had extended his analysis to Indigenous, feminist, and/or postcolonial fiction, which often challenge the human/nonhuman binary of western patriarchal thought. However, his analysis is particularly valuable for thinking about a history of silence and erasure when it comes to telling stories about the energies that undergird global capitalism—and, I'd add global militarism—in "the preserves of serious fiction".

In the time I have remaining I want to turn to Chamorro author Craig Santos Perez, who has written extensively about the voyaging canoe metaphor in the wake of transoceanic militarism, and might be the only poet on the planet to turn to the RIMPAC exercises and inscribe their impact on both human and nonhuman ocean ecologies. While his medium is experimental poetry rather than the realist novel, he challenges western binary thinking that separates the human from its nonhuman others and the separation of militarism from the transoceanic imaginary.

Author of a multi-book project entitled *from unincorporated territory*, a reference to the political status of Guahån, Perez is the winner of a PEN award and "imagines the blank page as an excerpted ocean, filled with vast currents, islands of voices, and profound depths."[32] Like other Indigenous poets from Oceania, a term Hau'ofa famously suggested as more representative of the flows of the region than the "Pacific," Perez has positioned his poetry as an oceanic vessel.[33] In his book covers, Perez juxtaposes images of Indigenous voyaging and fishing traditions (particularly in *[hacha]* and *[saina]*) over the aerial gaze preferred by US military photographs. This juxtaposition in terms of thematic, color, and vision between Chamorro and military life only changes in his latest collection, *[lukao]*, which gives us a submarine world as viewed by his infant daughter.

In only ten years of publishing, Perez has provided an ongoing critique of 21st century transoceanic militarism, and renders a military that is often hidden in plain sight. Since the beginning of his *from unincorporated territory* series, (*[hacha]* in 2008), Perez has critiqued the history and depiction of Guahån as a strategic naval base, as "USS *Guam*," and has framed his poems as "provid(ing) a strategic position for 'Guam' to emerge" from colonial and military hegemony. As such, he draws extensively on Indigenous voyaging traditions to poetically contest the US Navy, reshaping what Ghosh has called the "accustomed barques of narration." The cover of *from unincorporated territory [saina]* (2010) juxtaposes a drawing of a Chamorro voyaging canoe, or sakman, above a photograph of the aircraft carrier, USS *Abraham Lincoln*, leading smaller naval ships in their patrol of the Indian Ocean in 2008.[34]

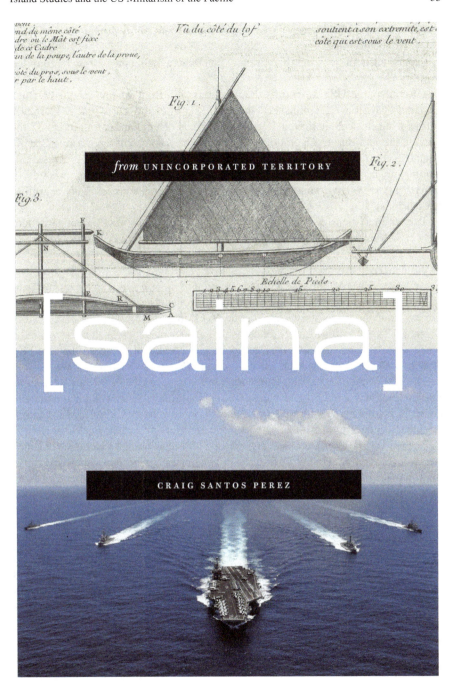

Although the world ocean has been partitioned into discrete national and international territories via the United Nations Law of the Sea (UNCLOS), the Navy considers each of its aircraft carriers "four and a half acres of sovereign US territory."[35]

Of course, the USS *Lincoln* is the infamous ship from which George W. Bush declared "Mission Accomplished" in May 2003 after the ship launched "16,500 sorties from its deck, and fired 1.6 million pounds of ordinance from its guns" the previous month during Operation Iraqi Freedom.[36] For complex reasons, Pacific Islanders continue to serve in disproportionate numbers in US military campaigns, lending nuance to the juxtaposition of these two different maritime vessels of sovereignty in which Chamorro claims are tied to Indigenous sovereignty as well as US patriotism.[37]

Perez's four books of poetry focus on US Naval colonialism in Oceania, particularly in Guåhan where it occupies one third of an island that is only 30 miles long.[38] In his most recent book, *from unincorporated territory [lukao]* (2017), he turns directly to the RIMPAC exercises of 2014. Thus Perez shifts the focus from the US Navy to the larger RIMPAC alliance, calling attention to the ways in which *transnational militarism* across the Indian and Pacific Oceans reflects a new era of hydro-politics. For example, the US military's publication "A Cooperative Strategy for 21st Century Seapower" emphasizes a closer relationship between agencies, such as between the Navy, Coast Guard, and Marines, as well as international alliances that are also evident in the Department of Defense's *Climate Change Adaptation Roadmap*. Both call for a new era of HADR or Humanitarian Assistance Disaster Relief operations because global warming is considered a threat multiplier.[39] It is well known that the commander of what was formerly the largest US Naval base (USPACOM), Admiral Locklear, declared climate change the biggest security threat to the Pacific in 2013.[40] Since then US Naval officers have argued for a "war plan orange for climate change" which involves more HADR operations in other countries because "these overtures may increase US access and these nations' receptiveness to hosting temporary basing or logistics hubs in support of future military operations."[41] Hence we have a call for larger RIMPAC activities, a 25% increase of ships to the Middle East, and a 60% increase in ships and aircraft by 2020 to the new ocean known the Pentagon as the "Indo-Asia-Pacific."[42] These are the military hydro-politics for the Anthropocene.

Perez's RIMPAC poem weaves together the fluid intimacy between mother and newborn daughter alongside the larger scale militarism of Oceania. The poem is entitled "(first ocean)" and its epigraph reads "*during the rim of the pacific military exercises, 2014.*" It is a second-person address to the poet's wife (you) and daughter (neni).Here is the poem in its entirety:

(first ocean)
during the rim of the pacific military exercises, 2014

~

when [neni] was newborn, [you] rinsed
her in the sink//pilot whales, deafened

by sonar, are bloated and stranded
ashore\\now [you] bathe her in the tub,

clean behind her ears, sing "my island
maui," written by your dad//his ashes

scattered in the pacific decades ago
\\when [we] bring [neni] to the beach

for the first time, [you] secure her
to your chest and walk into the sea\\

what will the aircrafts, ships, soldiers,
and weapons of 22 nations take from [us]

//"i wish she could've met my dad," [you] say\\
schools of recently spawned fish, lifeless,

spoil the tidelands//is oceania memorial
or target, economic zone or monument,

territory or mākua//a cold salt wind surges
\\[we] shiver like generations of coral reef
bleaching

The poem intersperses the Navy's ecological damage to all oceanic creatures—human and otherwise—with his newborn daughter's first immersion in the ocean. The use of parentheses in the poem's title invokes a placental or bodily enclosure of the infant, perhaps reminding the reader—like the conclusion of his eco-film/poem "Praise Song for Oceania"—that "our briny blood" connects us to the sea and our first placental ocean.[43]

The poet traces out the baby's first introductions to water by her mother in Hawai'i, moving from being rinsed in the sink, to a bath, to immersion in the sea. Each watery rinsing, bathing, and cleaning is juxtaposed to the repercussions of naval militarism: "pilot whales, deafened/by sonar" emerge "bloated and stranded/ashore". The speaker wonders "what will the aircrafts, ships, soldiers/and weapons of 22 nations take from [us]." In response, we learn of the loss of the child's grandfather, whose ashes were "scattered in the pacific decades ago," as well as the death of "schools of recently spawned fish" that lie in the tidelands, "lifeless". It has been widely reported that whale strandings and other animal deaths increase during and after RIMPAC exercises.[44] The poem concludes with a haunting question: "is Oceania memorial/or target, economic zone or monument/territory or mākua."

Mākua is the Hawaiian word for parent but also refers to the highly contested military reservation at Mākua Valley on Oʻahu, a place in Kanaka Maoli stories where humans originated, yet is now where sacred Hawaiian sites and endangered species have been regularly bombed since the 1920s.[45] The poem calls attention to the ways in which the militarization of Oceania causes a rupture in the responsibilities of the mākua to the child, a rupture in the kuleana or chain of responsibility that connects all living and beings and matter. The collection as a whole, by telescoping between the ordinary and the catastrophic, maternal intimacy and a militarized world ocean, brings together the very components that Ghosh notes are central to our understanding of the Anthropocene, yet so difficult to narrate in (western) Anglophone prose. Perez demonstrates the "uncanny intimacy of our relationship with the nonhuman" as Ghosh describes it, and raises vital questions about intergenerational survival and responsibility.

The 2014 RIMPAC "war games" invoked by the poem led to the widespread devastation of marine wildlife and a 2015 ruling by a Federal Judge that the US Navy exercises, especially the use of explosives and sonar, were endangering millions of marine mammals.[46] The court determined that there was a "breathtaking assertion" by the US Navy that their oceanic exercises "allow for no limitation at all," in terms of time, space, species, or depth, and that there was no justification for needing "continuous access to every single square mile of the Pacific."[47] Moreover, in a critical—if not cleverly literary—ruling, Federal Judge Susan Oki Mollway determined:

> Searching the administrative record's reams of pages for some explanation as to why the Navy's activities were authorized by the National Marine Fisheries Service ("NMFS"), this court feels like the sailor in Samuel Taylor Coleridge's "The Rime of the Ancient Mariner" who, trapped for days on a ship becalmed in the middle of the ocean, laments, "Water, water, every where, Nor any drop to drink."[48]

A critical ocean studies for island studies and the Anthropocene would bring together geopolitics with the literary and, like the poet Craig Santos Perez and Judge Mollway, narrate them in ways that mutually inflect and inform each other. While the recent oceanic turn has produced scholarship that presses our understanding of the ontological fluidity of our oceanic planet, a vigorous engagement with naval hydropolitics, a critical militarization studies, would help us better articulate and imagine a demilitarized future.

Endnotes

1. Brazil was invited but withdrew, reducing the number to twenty-five. China was "disinvited" due to its territorial expansion in the South China Sea. See "RIMPAC 2018 Begins."
2. "U.S. Navy Announces."
3. "U.S. Navy Announces."
4. Hayes, Zarsky, and Bello, *American Lake*. On China, see Forsythe, "Possible Radar" and the essays collected in Prabhakar, Ho, and Bateman, *The Evolving Maritime Balance*.

5. "Critical ocean studies" is explored in DeLoughrey, "Submarine Futures." This article is in conversation with important work by Steinberg and Peters, "Wet Ontologies"; Alaimo, *Exposed*; and Neimanis, *Bodies of Water*.
6. This is a larger argument taken up in relation to the British maritime (and shipwreck) fiction as well as more recent black Atlantic discourse in my *Routes and Roots*.
7. See Bélanger and Sigler, "Wet Matter."
8. Earle, *Sea Change*, xiv and NOAA biologist, Nancy Foster, quoted in Earle xiv.
9. The term was first used by Andreas Malm and then developed by Jason Moore and Donna Haraway. See Malm, *Fossil Capital*; Moore, *Capitalism in the Web of Life*; and Haraway, *Staying with the Trouble*.
10. Liska and Perrin, "Securing Foreign Oil."
11. See Secretary of the Navy J. William Middendorf III's 1974 address to the Rotary Club of San Francisco, who argued "It is the mission of the US Navy to protect the sea lanes for the transport of these critical [energy] imports. And it is the mission of the US Navy to render a political and diplomatic presence in the world today in support of our national policy." In Middendorf, "World Sea Power," 241.
12. Lawrence, "US Military is a Major Contributor"; Hynes, "Military Assault on Global Climate"; Sanders, *The Green Zone*. On the estimation of the number of US military bases (many of which are top secret) see Johnson, *Nemesis* and Lutz, *The Bases of Empire*. While Lutz calculates at least 1,000 overseas bases, the Department of Defense itself declares it has over "7,000 bases, installations, and other facilities" in its "2014 Climate Change Adaptation Roadmap".
13. Sanders, *The Green Zone*.
14. The Pentagon was given an exemption from reporting its carbon emissions at the Kyoto Convention on Climate Change. See Hynes, "Military Assault on Global Climate," and Neslen, "Pentagon to lose emissions exemption."
15. Hynes, "Military Assault on Global Climate."
16. Mahan, *The Influence of Sea Power*. On the history, see Thompson *Imperial Archipelago*.
17. See Adomeit, "Alfred and Theodore Go to Hawai'i." Mahalo to Anne Keala Kelly for this reference and her kokua regarding the naval history of Hawai'i.
18. The US Navy also ruled American Samoa from 1900 to 1951 which catalyzed the Mau protests. See David A. Chappell, "The Forgotten Mau: Anti-Navy Protest in American Samoa, 1920–1935," in *Pacific Historical Review*, Vol. 69, No. 2 (May, 2000), pp. 217–260. My thanks to my colleague Keith Camacho for his insights on US Naval rule in the Pacific Islands. On resistance to militarism in the Pacific see Shigematsu and Camacho, *Militarized Currents*. On the Indigenous responses to American and Japanese militarism in the Marianas see Camacho, *Cultures of Commemoration*, and Camacho, *Sacred Men*. See also Craig Santos Perez's preface to his first volume *from unincorporated territory [hacha]* for the Chamorro historical context.
19. On the weapons testing zones see Van Dyke "Military Exclusion and Warning Zones."

20. "United States Indo-Pacific Command."
21. "RIMPAC is the world's largest international maritime exercise." *U. S. Navy*, http://www.cpf.navy.mil/rimpac/2014/ (accessed August 3, 2018).
22. Klare, "Garrisoning the Global Gas Station."
23. Terrell, Hunt and Gosden, "The Dimensions of Social Life." This is discussed in DeLoughrey, *Routes and Roots*, 104–105.
24. Key texts that used the ocean as a trope for globalization include Connery, "The Oceanic Feeing," and Hau'ofa "Our Sea of Islands." These two edited collections were critical to shifting US literary and cultural studies to the Pacific.
25. Bauman's liquid metaphors for globalization (in *Liquid Modernity*) are discussed in DeLoughrey, *Routes and Roots,* 225–226.
26. Reich, "Trans-Pacific Partnership."
27. Wallach, "NAFTA on Steroids," and Solomon and Beachy, "A Dirty Deal."
28. Brooke, "Looking for Friendly Overseas Base." On "lily pads" and "forward operating locations" see Lutz, *The Bases of Empire*, 20, 37. See also Natividad and Kirk, "Fortress Guam."
29. See Norris, "The Air Assault on Japan," 86.
30. Steinberg, *The Social Construction of the Ocean.*
31. Hau'ofa, *We are the Ocean*; Jolly, "Imagining Oceania"; Clifford *Routes*. Diaz & Kauanui have argued that the "Pacific is on the move," understood in terms of tectonics, human migration, and a growing field of scholarship in "Native Pacific Cultural Studies," 317; I've built on these works in *Routes and Roots*, which makes an argument for a "transoceanic imaginary" (37).
32. "The Page Transformed."
33. Hauʻofa, *We are the Ocean.* I explore the oceanic vessel metaphor in greater depth in *Routes and Roots.*
34. Perez, *from unincorporated territory [saina].*
35. See Lutz, *The Bases of Empire*, 4.
36. Sanders, *The Green Zone*, 60.
37. See the introduction and essays collected in Bascara, Camacho, and DeLoughrey, "Gender and Sexual Politics."
38. See Camacho and Monnig, "Uncomfortable Fatigues," 158.
39. Mabus et al., "Cooperative Strategy for 21st Century Seapower," and Department of Defense, "2014 Climate Change Adaptation Roadmap."
40. Bender, "Chief of US Pacific forces."
41. McGeehan, "A War Orange for Climate Change."
42. Mabus et al., "Cooperative Strategy for 21st Century Seapower."
43. Earle, *Sea Change,* 15.
44. Fergusson, "Whales beware."
45. Activist groups such as Mālama Mākua and Earthjustice have brought the military to court to halt the bombing, at least for the time being. On the militarism of Hawai'i and Mākua in particular, see Anne Keala Kelly's powerful film, *Noho Hewa: The Wrongful Occupation of Hawai'i.* See also Keala Carter's 2018 update, "U. S. Army wants to resume live-fire training."
46. "Court Rules Navy Training in Pacific Violates Laws."

47. Conservation v. National Marine Fisheries, 6876 (Hawai'i District Court, 2013). https://earthjustice.org/sites/default/files/files/2013-12-16NAVY SonarComplaint.pdf.
48. Conservation v. National Marine Fisheries Civ. No. 13-00684 SOM/RLP; Natural Resources v. National Marine Fisheries, Civ. No. 14-00153 SOM/RLP (Hawai'i District Court, 2015). https://earthjustice.org/sites/default/files/files/2015-3-31%20Amended%20Order.pdf.

Acknowledgments I would like to thank RIIS, the Research Institute for Islands and Sustainability, particularly Drs Yoko Fujita and Ayano Gizano, for their kind invitation to participate in the 2019 Islands Studies Symposium and subsequent publication.

References

Adomeit, A.L. (2016.) Alfred and Theodore go to Hawai'i: the value of Hawai'i in the maritime strategic thought of Alfred Thayer Mahan. *The International Journal of Naval History,* 13(1). http://www.ijnhonline.org/2016/05/26/alfred-and-theodore-go-to-hawaii-the-value-of-hawaii-in-the-maritime-strategic-thought-of-alfred-thayer-mahan/#fn-1869-1.

Alaimo, S. (2016). *Exposed: environmental politics and pleasures in posthuman times.* Minneapolis: University of Minnesota Press.

Bascara, V., Camacho, K. L., & DeLoughrey, E. (Eds.). (2015). Gender and sexual politics in the Pacific Islands: a call for critical militarisation studies. *Special issue. Intersections: Gender and Sexuality in Asia and the Pacific, 37.* http://intersections.anu.edu.au/issue37_contents.htm.

Bauman, Z. (2000). *Liquid modernity.* Cambridge: Polity

Bélanger, P., & Jennifer, S. (Eds.). (2014). Wet matter. *Harvard Design Magazine,* 39 (2014). harvarddesignmagazine.org/issues/39.

Bender, B. (2013) Chief of U.S. Pacific forces calls climate biggest worry. *Boston Globe,* March 9, 2013. https://www.bostonglobe.com/news/nation/2013/03/09/admiral-samuel-locklear-commander-pacific-forces-warns-that-climate-change-top-threat/BHdPVCLrWEMxRe9IXJZcHL/story.html.

Brooke, J. (2004). Looking for friendly overseas base, Pentagon finds it already has one. *The New York Times,* April 7, 2004. https://www.nytimes.com/2004/04/07/us/looking-for-friendly-overseas-base-pentagon-finds-it-already-has-one.html.

Camacho, K. (2011). *Cultures of commemoration: the politics of war, memory and history in the Mariana Islands.* Honolulu: University of Hawai'i Press.

Camacho, K. (2019). *Sacred men: law, torture, and retribution in Guam.* Durham: Duke University Press.

Camacho, K. L., & Monnig, L. A. (2010). Uncomfortable Fatigues: Chamorro Soldiers, Gendered Identities, and the Question of Decolonization in Guam. In S. Shigematsu & K. L. Camacho (Eds.), *Militarized currents: toward a decolonized future in Asia and the Pacific* (pp. 147–180). Minneapolis: University of Minnesota Press.

Carter, K. (2018) Let's bomb this!' U.S. Army wants to resume live-fire training in sacred Hawaiian valley. *Intercontinental Cry.* https://intercontinentalcry.org/lets-bomb-this-us-army-wants-to-resume-live-fire-training-in-sacred-hawaiian-valley/.

Chappell, D. A. (2000). The Forgotten Mau: Anti-Navy Protest in American Samoa, 1920-1935. *Pacific Historical Review,* 69(2), 217–260. https://doi.org/10.2307/3641439.

Clifford, J. (1997). *Routes: travel and translation in the late 20th century.* Cambridge: Harvard University Press.

Connery, C.L. (1996). The oceanic feeling and the regional imaginary. In R. Wilson & Wimal D. (Eds.), *Global/local: cultural production and the transnational imaginary* (pp. 284–311). Durham: Duke University Press.

Court Rules Navy Training in Pacific Violates Laws Meant to Protect Whales, Sea Turtles. *EarthJustice*, April 1 2015. https://earthjustice.org/news/press/2015/court-rules-navy-training-in-pacific-violates-laws-meant-to-protect-whales-sea-turtles.

DeLoughrey, E. (2019). *Allegories of the anthropocene*. Durham: Duke University Press.

DeLoughrey, E. (2007). *Routes and roots: navigating Caribbean and Pacific Island literatures*. Honolulu: University of Hawai'i Press.

DeLoughrey, E. (2017). Submarine futures of the anthropocene. *Comparative Literature, 69*(1), 32–44. https://doi.org/10.1215/00104124-3794589.

Department of Defense. 2014 Climate change adaptation roadmap. Alexandria, VA: Department of Defense. https://www.acq.osd.mil/eie/downloads/CCARprint_wForward_e.pdf.

Diaz, V.M. and Kauanui, J.K. (2001). Native Pacific cultural studies on the edge. *The Contemporary Pacific, 13*(2):315–342. https://scholarspace.manoa.hawaii.edu/bitstream/handle/10125/13574/v13n2-315-342.pdf?sequence=1.

Earle, S. A., & Change, S. (1995). *A Message of the Oceans*. New York: Putnam.

Fergusson, M. (2014). Whales beware. In *The Hawai'i independent*, August 22, 2014. http://hawaiiindependent.net/story/whales-beware.

Forsythe, M. (2016). Possible Radar suggests Beijing wants 'effective control' in South China Sea. *The New York Times*, February 23, 2016. https://www.nytimes.com/2016/02/24/world/asia/china-south-china-sea-radar.html.

Ghosh, A. (2016). *The great derangement: climate change and the unthinkable*. Chicago and London: University of Chicago Press.

Ghosh, A. (1992). Petrofiction: the oil encounter and the novel. *New Republic, 9*(206), 29–34.

Ginoza, A. (2015). Dis/articulation of ethnic minority and indigeneity in the decolonial feminist and independence movements in Okinawa. *Intersections: Gender and Sexuality in Asia and the Pacific*, 37. http://intersections.anu.edu.au/issue37_contents.htm.

Haraway, D. J. (2016). *Staying with the Trouble: Making Kin in the Cthulucene*. Durham: Duke UP.

Hau'ofa, E. (1995). Our sea of islands. In Rob Wilson and Arif Dirlik (Eds.), *Asia/Pacific as Space of cultural production* (pp. 86–100). Durham: Duke University Press.

Hau'ofa, E. (2008). *We are the ocean: selected works*. Honolulu: University of Hawai'i Press.

Hayes, P., Zarsky, L., & Bello, W. (1986). *American lake: nuclear peril in the pacific*. New York: Viking.

Hynes, H. Patricia. (2011). The military assault on global climate. In *Truthout*, September 8, 2011. https://truthout.org/articles/the-military-assault-on-global-climate/.

Johnson, C. (2008). *Nemesis: the last days of the American Republic*. New York: Holt.

Jolly, M. (2001). Imagining Oceania: Indigenous and foreign representations of a sea of islands. In D. Yui & Y. Endo (Eds.), *Framing the Pacific in the 21st century: co-existence and friction* (pp. 29–48). Tokyo: Center for Pacific and American Studies, University of Tokyo.

Kelly, A.K. (2008). dir. Noho Hewa: The Wrongful Occupation of Hawai'i. Kailua, Hawai'i: Kuleana Works Production, DVD.

Klare, M.T. (2008). Garrisoning the global gas station. *Global Policy Forum*. https://www.globalpolicy.org/component/content/article/154-general/25938.html.

Lawrence, J. (2014). The U. S. military is a major contributor to global warming. *San Diego Free Press*, November 14, 2014. https://sandiegofreepress.org/2014/11/the-us-military-is-a-major-contributor-to-global-warming/.

Liska, A.J., & Perrin, R.K. (2010). Securing foreign oil: a case for including military operations in the climate change impact of fuels. *Environment: Science and Policy for Sustainable Development*. http://www.environmentmagazine.org/Archives/Back%20Issues/July-August%202010/securing-foreign-oil-full.html.

Lutz, C. (2009). *The Bases of Empire: The Global Struggle against U. S. Military Posts*. New York: NYU Press.

Mabus, Ray et al. "A Cooperative Strategy for 21st Century Seapower." U. S. Marine Corps, Navy, and Coast Guard, March 2015. http://www.navy.mil/local/maritime/150227-CS21R-Final.pdf.

Mahan, A. T. (1890). *The influence of sea power upon history 1660–1783* (p. 1987). New York: Dover.

Malm, A. (2016). *Fossil capital: the rise of steam power and the roots of global warming.* London: Verso.

McGeehan, T. (2017). A war plan orange for climate change. *U. S. Naval Institute Proceedings Magazine, 143*(10). https://www.usni.org/magazines/proceedings/2017-10/war-plan-orange-climate-change.

Middendorf, J.W. (1976). World sea power: U. S. vs. U. S. S. R. In II. William Menard & J. L. Schieber (Eds.), *Oceans: our continuing frontier.* Del Mar, CA: Publisher's Inc.

Moore, J. W. (2015). *Capitalism in the web of life: ecology and the accumulation of capital.* London: Verso.

Natividad, L., & Kirk, G. (2010). Fortress Guam: resistance to U. S. military mega-buildup. *The Asia-Pacific Journal, 8*(19):1–17. apjjf.org/-LisaLinda-Natividad/3356/article.html.

Neimanis, A. (2017). *Bodies of water: posthuman feminist phenomenology.* London: Bloomsbury.

Neslen, A. (2015). Pentagon to lose emissions exemption under Paris climate deal. *The Guardian*, December 14, 2015. https://www.theguardian.com/environment/2015/dec/14/pentagon-to-lose-emissions-exemption-under-paris-climate-deal.

Norris, J. G. (1943). The Air Assault on Japan. *Flying, 21–23,* 86.

Perez, C.S. (2010). The page transformed: a conversation with Craig Santos Perez. *Lantern Review Blog*, March 12, 2010. http://www.lanternreview.com/blog/2010/03/12/the-page-transformed-a-conversation-with-craig-santos-perez/.

Perez, C.S. (2016). Chanting the water. http://craigsantosperez.com/chanting-waters-2016/.

Perez, C.S. (2008). *From unincorporated territory [hacha].* Kaneohe, HI: Tinfish.

Perez, C.S. (2010). *From unincorporated territory [saina].* Oakland: Omnidawn.

Perez, C. (2017). *From unincorporated territory [lukao].* Oakland: Omnidawn.

Perez, C.S., & Chong, J.A. (2017). Praise song for Oceania. YouTube video, 4:56, June 14, 2017. http://craigsantosperez.com/praise-song-oceania/.

Povinelli, E. A. (2016). *Geontologies: a requiem to late liberalism.* Durham: Duke University Press.

Prabhakar, L. W., Ho, J. H., & Bateman, S. (2006). *The evolving maritime balance of power in the Asia-Pacific: maritime doctrines and nuclear weapons at sea.* Singapore: Institute of Defence and Strategic Studies.

Reich, R. (2015). Robert Reich on trans-pacific partnership as 'NAFTA on Steroids.' YouTube video, 2:26, posted by "Business and Human Rights Resource Centre," January 29, 2015, https://www.business-humanrights.org/en/robert-reich-on-trans-pacific-partnership-as-nafta-on-steroids.

RIMPAC 2018 Begins, But Without China. *The Maritime Executive.* June 29, 2018. https://www.maritime-executive.com/article/rimpac-2018-begins-but-without-china#gs.hahJMZ8.

Sanders, B. (2009). *The green zone: the environmental costs of militarism.* Chico, CA: AK Press.

Shigematsu, S., & Camacho, K. L. (Eds.). (2010). *Militarized currents: toward a decolonized future in Asia and the Pacific.* University of Minnesota Press: Minneapolis.

Solomon, I., & Ben, B. (2015). A dirty deal: how the trans-pacific partnership threatens our climate. Washington, DC: The Sierra Club. https://www.sierraclub.org/sites/, www.sierraclub.org/files/uploads-wysiwig/dirty-deal.pdf.

Steinberg, P. (2001). *The social construction of the ocean.* Cambridge: Cambridge University Press.

Steinberg, P., & Peters, K. (2015). Wet ontologies, fluid spaces: giving depth to volume through oceanic thinking. *Environment and Planning D: Society and Space, 33*(2), 247–264. https://doi.org/10.1068/d14148p.

Terrell, J.E., Hunt, T.L., & Gosden, C. (1997). The dimensions of social life in the Pacific: human diversity and the myth of the primitive isolate. *Current Anthropology, 38*(2):155–195. https://doi.org/10.1086/204604.

Thompson, L. (2010). *Imperial Archipelago: representation and rule in the insular territories under U. S. Dominion after 1898.* Honolulu: University of Hawai'i Press.

United States Indo-Pacific Command. *Wikipedia*. https://en.wikipedia.org/w/index.php?title=United_States_Indo-Pacific_Command&oldid=851229307. Accessed August 3, 2018.

U.S. Navy Announces 26th Rim of the Pacific Exercise. *Navy News Service*. May 30, 2018. http://www.navy.mil/submit/display.asp?story_id=105789.

Van Dyke, J.M. (1993). Military exclusion and warning zones on the high seas. In J.M. Van Dyke, D. Zaelke, & G. Hewison (Eds.), *Freedom for the seas in the 21st century: ocean governance and environmental harmony* (pp. 445–70). Washington, D.C.: Island P, 1993.

Wallach, L. (2012). NAFTA on steroids. *The Nation*, June 27, 2012. https://www.thenation.com/article/nafta-steroids/.

Island Studies Inside (and Outside) of the Academy: The State of this Interdisciplinary Field

James E. Randall

Introduction

It is difficult to estimate precisely when the interdisciplinary field of Island Studies had its start. Was it when the Institute of Island Studies was created at the University of Prince Edward Island in 1988? Or did it originate with the establishment of the International Small Islands Studies Association (ISISA) in 1992. Or perhaps the genesis of Island Studies can be traced to recognition by the United Nations of the Small Island Developing States (or SIDS) and the subsequent establishment of the Alliance of Small Island States (AOSIS), also in the early 1990s. Some might argue that Island Studies started when the first academic degree programs dedicated to Island Studies were established at the University of Malta in the early 1990s. Symbolically, others would point to Grant McCall coining the term *nissology* to describe "the study of islands on their own terms" (McCall 1994).

Regardless of the precise action or date, Island Studies has now been a part of our lexicon for at least one-quarter of a century. During that time, many researchers have reflected on the state and direction of scholarship in island studies (Baldacchino 2004; Fletcher 2011; Grydehøj 2017). However, there has been less attention paid to the evolution and future of what might be referred to as the "institutional framework" of Island Studies; in other words, how and where it is carried out and by whom. This is important because the functions of research, teaching, public engagement and advocacy are created and recreated largely through established institutional structures. Therefore, the purpose of this paper is to describe and evaluate the institutional administrative structures associated with island studies; to take stock of how and where island studies is practiced as a field of enquiry and where it might be headed in the future.

J. E. Randall (✉)
University of Prince Edward Island, Charlottetown, Canada
e-mail: jarandall@upei.ca

The Institutions of Island Studies

There are now many island-based institutions or organisations around the world that are practicing "island studies", whether this occurs at the scale of a single island or archipelago, a larger region of islands (e.g., the Pacific) or encompasses the entire "world of islands" (Robertson 2018). These institutions of island studies can be grouped into several complementary and sometimes intersecting categories. First, there are those supranational or intergovernmental organizations that undertake a combination of advocacy, research, policy development and/or public engagement at a global or regional scale. Second, there are island-focused research centres/institutes or "think tanks", often dedicated to examining issues related to island sustainability or development. Third, there is a related group of institutes that are more closely affiliated with universities or colleges and therefore often incorporate an academic training dimension into their activities. Finally, there are "communities" of individuals who share a common interest and passion for islands and who organize themselves in the form of professional associations, publish in scholarly journals and meet at conferences, established for the purpose of disseminating island studies research.

Supranational Island Organisations

There are a large number of international bodies that represent islands in one form or another on the world stage. It is not uncommon for these to be directly or indirectly affiliated with island governments. This includes the aforementioned Alliance of Small Island States, the Organization of Eastern Caribbean States (OECS), the Secretariat of the Pacific Community (SPC), the European Small Islands Federation or network (ESIN), the CPMR European Islands Commission, and the B7 Baltic Islands Network. The primary activity of these organizations appears to be advocating or lobbying for some form of change that will benefit their member islands politically and financially. In terms of the number and maturity of these institutions, the European region of islands (i.e., Mediterranean, Baltic and the North Atlantic) seems to be further advanced that any other area. For example, the European Small Islands Federation (ESIN) is an umbrella group that represents the interests of eleven separate small islands federations throughout Europe. Collectively, they represent the "…voice of 359,357 islanders on 1,640 small islands, helping them remain alive."[1] At the level of the European Union, the CPMR European Islands Commission advocates to influence policy and rules in favour of small islands. At a local level, they try to strengthen cultural identity and share knowledge across the region on critical issues. Their aim is, "To urge the European Institutions and Member States to pay special

---

[1] From https://europeansmallislands.com/ (accessed February 19, 2019).

attention to the islands, to acknowledge the permanent handicaps resulting from their insularity, and to implement policies that are best suited to their condition."[2]

Outside of Europe, but still with a European connection, the OCTA group (Association of the Overseas Countries and Territories of the European Union) consists of 22 members drawn from among the subnational island jurisdictions affiliated with France, Britain and the Kingdom of the Netherlands. It has a mandate to promote common positions and partnerships for the sustainable development of their members through cooperation, capacity building and communication. The 44-member Association of Small Island States has an explicit mandate to advocate for small island states on a global stage and attempt to shape international environmental policy. This organisation has been most closely associated with advocating for global climate change mitigation policies and strategies, for example at the COP21 United Nations Framework Convention on Climate Change held in Paris in December, 2015 (Ourbak and Magnon 2018).

As these examples show, the mandates of some of these institutions speaks to a broader overarching goal of improving the well-being of islanders within their jurisdictions, whether this is done through lobbying, research or the exchange of people and products. For example, the vision of the B7 Baltic Islands Network is to make island life more attractive and sustainable by exchanging experience, lobbying and projects. In addition, not all of these organisations are affiliated with independent island states. They can also represent subnational territories or dependencies such as is the case with OCTA. Alternatively, with the B7 Baltic group, they can represent more local, municipally-based island jurisdictions. For example, all of the current members of the B7 Baltic Islands Network are island municipalities, states or provinces (i.e., subnational island jurisdictions) including the Åland islands (Finland), Gotland (Sweden), Hiiumaa (Estonia) and Rügen (Germany).[3]

Some island-based institutions consist of members drawn from a broader range of stakeholders, including foundations and NGOs (non-governmental organisations). One of the most prominent in terms of its internet presence is the **GL**obal **IS**land **PA**rtnership (GLISPA). The mission of GLISPA is to promote action to build resilient and sustainable island communities by inspiring leadership, catalyzing commitments and facilitating collaboration for all islands.[4] Started in 2006 by the Presidents of Palau and Seychelles, it is co-chaired by these two Presidents and the Prime Minister of Grenada. It currently has 40 members and claims to have "catalyzed" $150 million US for island action, although it is not clear how this is measured or the precise role GLISPA has played in raising these funds. The high-level political presence associated with GLISPA appears to have enhanced the reputation of the organization. It is also one of the few island organisations to charge an annual membership fee, starting at $5,000 US.[5]

---

[2] From https://cpmr-islands.org/who-we-are/ (accessed February 19, 2019).
[3] The Åland islands is also home to the Åland International Institute for Comparative Island Studies.
[4] From http://www.glispa.org/ (accessed February 19, 2019).
[5] Affiliate members are charged between $500 and $4,999 and "Friends" are not charged a memebrships fee.

A more recent entry into the arena of island studies institutional networks is the consortium known by the acronym RETI, which stands for Reseaux d'excellence des Territoires Insulaires (in English, the Excellence Network of Island Universities).[6] Established in 2010 and administered by the University of Corsica, it consists of 27 member universities or colleges spanning all regions of the world. Although not many of the RETI universities have Island Studies Departments or Programs, they all share the characteristic that their universities play significant roles on the islands on which they are located, not only by offering educational programs for islanders but also by their ability to influence the social, economic, cultural and environmental development of the islands upon which they are located. The RETI group holds annual meetings consisting of three components; a traditional conference where the faculty and graduate students from the member universities can present their research, a short training session for students, and a meeting of the University Presidents/Rectors/Principals or their designates to discuss common challenges. There has been increased interest in this group recently by Asian island universities and by universities located on subnational island jurisdictions, including the addition of the National Penghu University, on Penghu, Taiwan, and the University of the Ryukyus, on Okinawa, Japan. The most recent meeting of RETI took place on the Canary Islands in early May, 2019.

Island Research Centres or Institutes

There are a growing number of island-focused institutions that might be referred to as research centres, institutes or "think-tanks". Many of these are dedicated to examining single issues, such as biodiversity (e.g., the Small Islands Organisation or SMILO), management of natural resources (e.g., CANARI—the Caribbean Natural Resources Institute), or renewable energy (e.g., SIDS DOCK, referring to small island developing states as global docking stations).[7] Although these organisations play crucial roles in addressing issues that are extremely relevant for small islands, this paper does not address them. We are more interested in those institutions that take a broader, integrative approach to island studies.

Island studies research institutes operate at various jurisdictional scales. In some cases their mandates are primarily associated with only one or a small number of islands. The example of Maine's (USA) Island Institute is a good example. It "works to sustain Maine's island and coastal communities, and exchanges ideas and experiences to further the sustainability of communities" in Maine and elsewhere.[8] Established in 1983, it explicitly links its activities to the inhabitants of the 120 island and coastal communities of the State of Maine, through education, outreach and capacity

---

[6] From https://reti.universita.corsica/ (accessed February 19, 2019).
[7] The website of CANARI is at https://www.canari.org/; the SIDS DOCK site is https://sidsdock.org/ and SMILO is found at http://www.smilo-program.org/en/ (accessed February 19, 2019).
[8] From http://www.islandinstitute.org/ (accessed February 19, 2019).

building. It is noteworthy that even in cases such as this very locally grounded institution, there is an increasing trend to reach out to other similar island communities. For example, Maine's Island Institute makes reference to the 150,000 year-round islanders located throughout the continental USA. Another example of this tendency to look both at the local island context and more broadly at other islands may be reflected in the renaming of this organization at the University of the Ryukyus; from the International Institute for Okinawan Studies to the Research Institute for Islands and Sustainability (RIIS). This renaming makes it clearer that it is concerned not only with sustainable development on Okinawa and Japan but elsewhere in the world as well.

University-Based Island Institutes

The more prominent international island studies research centres or institutes appear to be grounded in a local island context but explicitly take this broader, global and comparative islands approach. Two of these organisations deserve special mention: the Institute of Island Studies (IIS) at the University of Prince Edward Island in Canada and the University of Malta's Islands and Small States Institute (ISSI).

Established in 1988, the IIS has always played a prominent role on its own island of 155,000 as well as within the global community of islands.[9] This includes the establishment of a federally-funded Canada Research Chair in Island Studies held for ten years by Godfrey Baldacchino, hosting and administering the flagship open-access, peer-reviewed, online *Island Studies Journal*, and founding the Master of Arts in Island Studies post-graduate academic program. After restructuring in 2013 and being awarded a UNESCO Chair in Island Studies and Sustainability in 2016, the IIS has positioned itself as an organization committed to building capacity in the study of island issues on PEI and elsewhere in the "sea of islands".[10] Its vision is "to be the leading centre of excellence on issues related to island studies scholarship, public policy, and engagement."[11]

Starting at about the same time as Prince Edward Island's Institute of Island Studies, the Islands and Small States Institute at the University of Malta has also taken a broad, interdisciplinary approach to the study of islands, promoting research and training on economic, social, cultural, ecological and geographical aspects of islands and small states.[12] Among other activities, the ISSI offers a Master's degree program/course in Islands and Small States Studies as well as what might be the only

---

[9] See http://projects.upei.ca/iis/ (accessed February 19, 2019).

[10] The UNESCO Chair is jointly held by Drs. James Randall at UPEI and Godfrey Baldacchino, now back at the University of Malta. It should also be noted that the growth of island studies and many of the institutions discussed in this paper would not have been possible without the leadership of Godfrey Baldacchino. Island studies scholars and practitioners owe him a debt of gratitude.

[11] The vision and mandate of the Institute of Island Studies are found at http://projects.upei.ca/iis/about-us/.

[12] From https://www.um.edu.mt/islands (accessed February 19, 2019).

Doctoral-level program in Island Studies. Although by no means the only universities to offer post-graduate degrees in Island Studies, UPEI and the University of Malta have been at the forefront in developing a community of graduates who have been trained in the scholarship and research methods associated with this interdisciplinary field.

Perhaps without the same capacity as UPEI's IIS and the UofMalta's ISSI, the University of the Highlands and Islands (UHI) in Scotland has established a strong presence across the archipelagos in northern and western Scotland, including the Orkneys, the Shetlands, and the Hebrides. Through their Institute for Northern Studies, they offer a Master of Letters (MLitt) in Island Studies using video technology.[13] Also in this category is the University of the South Pacific (USP) with its Faculty of Islands and Oceans. Unlike the universities noted above, this Faculty at the USP does not explicitly include a Department or degree program in Island Studies. However, all of the Departments and Programs under this Faculty are expected to embody their vision to be "a world-renowned centre of excellence in terms of sustainable management and development of our island, ocean and human resources" and share a mission to "respond to the needs of Pacific Island communities, through relevant teaching and learning, research, community outreach and other services."[14] It is noteworthy that the first guiding principle of this Faculty of Islands and Oceans is to "Think Pacific".

There are a number of other university academic programs at Pacific universities that are focused on islands in the Pacific region. Kagoshima University's Research Center for the Pacific Islands has operated in its present form since 1998.[15] Its objective is to promote interdisciplinary studies on islands and island zones in Oceania and its surroundings. The University of Hawaii's Center for Pacific Island Studies on the Mānoa campus is both an academic department and a research/public engagement centre. It offers a Master of Arts, a Bachelor of Arts and a Graduate Certificate, all in Pacific Islands studies. However, unlike the situation at UPEI, the UofMalta, or UHI, the Center for Pacific Island Studies does not take a global or an interdisciplinary perspective. Their academic programs focus on one geographic region and thematically on research and learning that relates specifically to cultural studies and post-colonialism. The University of Hawai'i is only one of 14 universities in the Pacific offering some form of Pacific Studies degree, including the University of Guam's Bachelor of Arts in Pacific-Asian Studies.

This description of some of the island studies degree programs is most noteworthy for the relative paucity of offerings. Despite the estimated 600 million people who live on islands throughout the world, there are relatively few university or college degree programs devoted to this area of enquiry. For comparison, take for example the presence of other area studies degree programs at the undergraduate and post-graduate levels. In the United States there are 91 Bachelor's level university or college level degree programs specializing in Asian Studies and 33 Master's level Asian Studies

---

[13] See https://www.uhi.ac.uk/en/courses/mlitt-island-studies/ (accessed February 19, 2019).

[14] From https://www.usp.ac.fj/index.php?id=5219 (accessed February 19, 2019).

[15] From http://cpi.kagoshima-u.ac.jp/introduction/overview.html (accessed February 22, 2019).

programs across North America and Europe. There are even eight degree programs at American colleges and universities that specialize in Scandinavian Studies, despite the fact that the population in all four Scandinavian countries accounts for just over 20 million people, or less than 4% of the world's island population.

The list of new institutional entries to island studies is growing. For example, the University of Exeter in England delivers and accredits a Master of Science degree in Island Biodiversity and Conservation on the Island of Jersey, one of the Channel Islands. They are exploring developing similar island-focused post-graduate degrees in island history and archaeology.[16] In collaboration with a Belgian university, the University of Aruba is establishing post-graduate degrees in SISSTEM (Sustainable Island Studies in Science, Technology, Engineering and Mathematics). And a new "think-tank" called the "Think To Do Institute" has been established on the island of Curaçao. The mission of this institute is to mobilize expertise and ideas to influence the policy making process on Curaçao. Finally, there are several proposals seeking funding from the European Union's Erasmus programs with the intention of establishing collaborative post-graduate degree programs in Island Studies from among groups of European universities.

Institutions of Island Scholars

When you look beyond the formal administrative units that specialize in island studies, you realize that there are many "communities of scholars" practicing research, teaching and engaging within their island communities who may not be affiliated with any of these organisations. These are the professional associations of academics, researchers and island practitioners who interact primarily through periodic meetings, conferences and symposia. The International Small Islands Studies Association (or ISISA) is the most well-known of these groups. The objectives of ISISA are, "…to study islands on their own terms, and to encourage free scholarly discussion on small island related matters such as islandness, smallness, insularity, dependency, resource management and environment, and the nature of island life."[17] Baldacchino (2018) notes that the main role of ISISA and its 100+ members is to plan for and organize upcoming conferences. It is noteworthy that the first official meeting of ISISA was held on Okinawa in 1994. It has held 16 "Islands of the World" meetings, with the most recent taking place in Leeuwarden and Terschelling, Netherlands. Although attendees at these conferences consist largely of faculty members and students at universities throughout the world, there is often a strong presence of local islanders and island organizations and an explicit attempt to provide value to the host islands and islanders.

The other longstanding community of scholars is the SICRI network. Established in 2004, SICRI stands for the Small Island Cultures Research Initiative. Its principal

---

[16]From personal correspondence with the Program director, Dr. Sean Dettman.

[17]From http://www.isisa.org/index.php?c=isisa-objectives (accessed February 19, 2019).

goal is to, "...research and assist the maintenance and development of the language, literature, music, dance, folkloric and media cultures of small island communities."[18] It has held 14 conferences, with the most recent taking place on Ile Tatihou, off the coast of Normandy, France. A more local example of these communities of academics and practitioners is the non-profit Japan Society of Island Studies. Their English language website states that they have more than 200 members who are interested or fascinated in islands.[19] They proudly note that more than half of their members are not academics.

Expressions of Island Studies Through Journals

One of the main ways in which the scholarship within academic fields is created, built, disseminated and critiqued is through the research published in scholarly journals. In Island Studies, there are two primary outlets for peer-reviewed research. The *Island Studies Journal* (ISJ) is considered the flagship periodical for peer-reviewed research, covering all aspects of island studies.[20] Established in 2006 and hosted by the University of Prince Edward Island, the ISJ had published 259 research papers and review essays up to the November 2018 issue. The Journal *SHIMA* publishes research on social, cultural, environmental and conceptual aspects of various types of islands.[21] It has published 214 articles or review essays since 2007. There are a number of other journals emerging that publish original research on more thematically focused island subjects. These include *Urban Island Studies*, started in 2015 and published under the Island Dynamics organization, the *Journal of Marine and Island Cultures* (published since 2012), *The Contemporary Pacific* and a new *Small States and Territories Journal* administered out of the University of Malta. Island Dynamics deserves special mention.[22] Created and coordinated by Adam Grydehøj (who is also the current Editor of the Island Studies Journal), Island Dynamics is perhaps best known for the conferences it coordinates. As of mid-February 2019, their website is promoting ten upcoming conferences, many of which are related to island contexts (e.g., Silk Road Archipelagos in Fuzhou, China; Asian-Arctic Connections in Nuuk, Greenland). The website also shows that Island Dynamics has organized 20 conferences since 2009 on topics such as "Investing in Small Island Recovery", "Island Cities and Urban Archipelagos", "Mermaids, Maritime Folklore, and Modernity", and "Islands, Resources and Society". Many of these are in non-Western settings, including Macau, Hong Kong, and Taiwan.

---

[18] From https://sicri-network.org/ (accessed February 19, 2019). The driving force behind SICRI has been Dr. Philip Hayward.

[19] From http://islandstudies.jp/jsis/ (accessed February 19, 2019).

[20] All full text articles associated with the *Island Studies Journal* are available free of charge at their website (https://www.islandstudies.ca/).

[21] From http://shimajournal.org/introduction.php.

[22] From https://www.islanddynamics.org/.

Across all of these institutions and organisations, there seems to be a growing awareness of what is taking place on other small islands. This goes beyond disseminating research in open-access journals and at conferences. Island Studies institutions are now more likely to know what other island organisations are doing, and are more willing to reach out to form new and stronger networks. This in itself bodes well for the future of island studies regardless of how it is practiced.

Discussion and Conclusions

This description of the institutions of island studies has undoubtedly missed many important organisations, especially those operating very locally, those that do not communicate their activities in English, and those that do not have a web presence. The issue of language is especially important. Not only does this review miss important institutions and outlets of scholarly island studies research that are published in languages other than English but, like the scholarship of island studies itself, it under represents research by indigenous islanders and from broad geographic regions with thousands of inhabited islands, including across South and Southeast Asia, Russia, and the near-shore islands of Africa (see Baldacchino 2008). Baldacchino goes on to note that, "The smaller, poorer or less populated the island gets, the more likely is it that its web, textual and literary content is dictated and penned by 'others'." (Baldacchino 2008, 38) The same criticism could be said for the academic programs and institutions associated with island studies. As is apparent from the examples used in this paper, many of these institutions and academic programs are dominated by the English language and the Western, developed world context.

The other feature of island studies is the absence of doctoral-level island studies programs at universities. Other than the few graduates from the University of Malta's PhD program in Island and Small State Studies, everyone affiliated with island studies institutions, conferences, research and academic programs has received their terminal degree in a discipline other than island studies. Most are trained in other social sciences or, less so, in the humanities. For example, of the contributors to the most recent compendium of island studies research, the *2018 Routledge Handbook of Island Studies*, nine of the authors are trained or teach in Geography, five are in Environmental Studies/Sciences Departments, four each are in Biology/Botany and Political Sciences/Policy Studies, and then there is a smattering of sociologists, historians, english/literary specialists, tourism specialists, economists, etc. I personally received all of my degrees in Geography. Perhaps more telling is the example of the University of Tasmania, Australia. It arguably has the largest contingent of English-speaking island studies scholars, each of whom have made significant contributions to the field of island studies.[23] However, none of these scholars were trained in doctoral-level Island Studies programs. In fact, the University of Tasmania does

---

[23] Drs. Lisa Fletcher, Andrew Harwood, Peter Hay, and Elaine Stratford.

not even offer an Island Studies degree program. This is important because postsecondary educational institutions may be one of the most powerful structures for development and representation of the ways we perceive and understand the world around us. If there truly is going to be a future for Island Studies as a stand-alone field of enquiry, there must be more doctoral-level programs dedicated specifically to Island Studies, graduating students who will eventually replace their mentors and supervisors in those same Island Studies programs.

These barriers to institutional entry in academia are not unusual for new fields of enquiry. As Soulé and Press (1998) note, "New ways of organizing teaching and research are often perceived as radical and threatening, particularly if there are fiscal implications for preexisting departments." (p. 401). The same could be said for other fields that have emerged at universities and in supranational organisations in the latter half of the twentieth century, most notably Environmental Studies, and Women's/Feminist/Gender Studies. It could be argued that these other areas of enquiry emerged because of the perception of an existential crisis in the world that could not be addressed through the existing institutional structures. In the case of Environmental Studies, the field may have started with the publication of Rachel Carson's seminal *Silent Spring* (1962), and the ensuing concern regarding the impacts of humankind on the state of the planet's environment (Cooke and Vermaire 2015; Soulé and Press 1998). Increasing student demand then forced universities to respond by developing curricula that could respond to this social agenda. In the case of women's and gender studies, the subjugation and absence of women from intellectual enquiry may have been the existential crisis that precipitated the rise in women's liberation movements and the subsequent explosion of courses and academic programs in this field (Boxer 1982; Stimpson 1973; Tobias 1978).

So is there an existential crisis that we can point to that has resulted in the growth of island studies institutional structures and may be a source of continued attention? Part of this "crisis" might be linked to the growing geopolitical importance of islands and the oceans surrounding them. There are at least three factors that have contributed to this growing prominence. First, with the implementation of the United Nation's Convention on the Law of the Sea that came into effect in 1994, island states and territories have gained formal control and responsibility over vast marine areas surrounding their islands and archipelagos. This has already had a major impact on the economic future of small islands and the relations they have with other states, companies and supranational organisations. It has also forced former imperial states such as France, Britain, the Kingdom of the Netherlands and the United States to reassess their relationships with their own subnational island jurisdictions, including injecting more resources into these islands (Fisher 2011).

Second, over the past seventy years, the proportion of United Nations (UN) members that are islands has been steadily increasing. When the United Nations was formed in 1945 with 51 members, only six (11.8%) were islands. By 2015, the number of UN members had grown to 193 and 45 of these (or 23.3%), were either a single island or, more commonly, groups of islands (Watts 2009). This has undoubtedly contributed to the growing voice of small island governments on global issues such as human-induced climate change and sustainable development.

Third, as a major power in the 21st century, China has taken a renewed interest in islands and ocean states. This has been reflected in growing partnerships and investments with island states along the Maritime Silk Road, but also in aggressive territorial claims of islands and ocean spaces in the South and East China Sea (Baldacchino 2016; Djankov and Miner 2016) and in the Arctic. In many respects, growing geopolitical tensions, adverse impacts of climate change and in particular rising sea levels, and development challenges facing small islands today may be the three-pronged existential crisis that will contribute to growth in island studies institutions. This parallels that which has taken place in the Arctic over the past generation, where larger mainland states have been jockeying with each other to exercise their territorial claims (Brutschin and Schubert 2016). In doing so, they have invested significant resources in their own Arctic infrastructures and institutions, including representation in supranational institutions such as the University of the Arctic and lobbying for observer status in the Arctic Council as the preeminent intergovernmental regime in the region (Gamble and Shadian 2017; Olsen and Shadian 2016). Although this renewed attention on islands by other geopolitical powers can and has led to abuse and exploitation, it can also provide resources for the development of new and more powerful island studies institutions. All of these factors also point to the need for a growing presence in island studies institutions and a better coordination and collaboration among them in order to best serve islands and islanders.

References

Baldacchino, G. (2004). The coming of age of Island studies. *Tijdschrift Voor Economische en Sociale Geografie, 95*(3), 272–283.
Baldacchino, G. (2008). Studying islands: On whose terms? Some epistemological and methodological challenges to the pursuit of Island studies. *Island Studies Journal, 3*(1), 37–56.
Baldacchino, G. (2016). Diaoyu Dao, Diaoyutai or Senkaku? Creative solutions to a festering dispute in the East China Sea from an 'Island Studies' perspective. *Asia Pacific Viewpoint, 57*(1), 16–26.
Baldacchino, G. (Ed.). (2018). *The Routledge international handbook of island studies: A world of Islands*. Routledge.
Boxer, M. J. (1982). For and about women: The theory and practice of women's studies in the United States. *Signs: Journal of Women in Culture and Society, 7*(3), 661–695.
Brutschin, E., & Schubert, S. R. (2016). Icy waters, hot tempers, and high stakes: Geopolitics and geoeconomics of the Arctic. *Energy Research and Social Science, 16,* 147–159.
Carson, R. (1962). *Silent spring*. London: Hamish Hamilton.
Cooke, S. J., & Vermaire, J. C. (2015). Environmental studies and environmental science today: Inevitable mission creep and integration in action-oriented transdisciplinary areas of inquiry, training and practice. *Journal of Environmental Studies and Sciences, 5*(1), 70–78.
Djankov, S., & Miner, S. (Eds.). (2016). *China's belt and road initiative: Motives, scope, and challenges*. Peterson Institute for International Economics.
Fisher, D. (2011). France in the South Pacific: An Australian perspective. In B. Neilson & R. Aldrich. (Eds.), *French history and civilization: Papers from the George Rudé seminar* (pp. 237–254). Accessed at http://www.h-france.net/rude/rudepapers.html.
Fletcher, L. (2011). '… some distance to go': A critical survey of Island Studies. [Paper in special issue: The Literature of Postcolonial Islands. DeLoughrey, E. (Ed.).]. *New Literatures Review,* (47–48), 17–34.

Gamble, J., & Shadian, J. M. (2017). One Arctic … But uneven capacity: The Arctic Council permanent participants. *ONE ARCTIC*, 142.

Grydehøj, A. (2017). A future of island studies. *Island Studies Journal, 12*(1), 3–16.

McCall, G. (1994). Nissology: The study of islands. *Journal of the Pacific Society, 17*(2–3), 1–14.

Olsen, I. H., & Shadian, J. M. (2016). Greenland and the Arctic Council: Subnational regions in a time of Arctic Westphalianisation. *Arctic Yearbook, 5,* 229–250.

Ourbak, T., & Magnan, A. K. (2018). The Paris agreement and climate change negotiations: Small islands, big players. *Regional Environmental Change, 18*(8), 2201–2207.

Robertson, G. (2018). Futures: Green and blue. In Baldacchino, G. (Ed.), *The Routledge international handbook of island studies: A world of islands* (pp. 416–441). Routledge.

Soulé, M. E., & Press, D. (1998). What is environmental studies? *BioScience, 48*(5), 397–405.

Stimpson, C. R. (1973). The new feminism and women's studies. *Change: The Magazine of Higher Learning, 5*(7), 43–48.

Tobias, S. (1978). Women's studies: Its origins, its organization and its prospects. *Women's Studies International Quarterly, 1*(1), 85–97.

Watts, R. (2009). Island jurisdictions in comparative constitutional perspective. In G. Baldacchino & D. Milne (Eds.), *The case for non-sovereignty: Lessons from sub-national Island jurisdictions* (pp. 21–39). London: Routledge.

The Perspective of Cultural Heritage/Cultural Landscape in Critical Island Studies

So Hatano

Introduction

It has been critically discussed by many researchers to date that the characteristics attributed to island space such as "remoteness" and "isolation" have been defined and physically shaped by colonial discourses. To overcome such definition based on modern territorial concepts, island space needs to be viewed as "an emergent product of relations" (Massey 2005: 68) arising from diverse encounters (which may be consonant or dissonant).

On the other hand, actions associated with the designation of cultural properties, which specify the cultural and historical values of the island landscape or items located in the area, lead to identification of the ideal state of the island through normative science. Furthermore, promotion of island tourism often takes the form of spatial development that presents a certain image, where tourists visit to consume that image. While these actions and development play a certain role in visualizing the value of the island (but it also means that value is sometimes imported from the outside), they also carry the risk of abandoning the multiplicity and plurality of value and oversimplifying the "relations" as discussed by Massey. Is it impossible, then, for normatization and multiplicity/plurality of value to coexist within the same island space?

Jinguashi Mines located in the northern part of Taiwan (ROC), an island in East Asia, was developed during the period of Japanese occupation. Its mining heritage may be characterized as a colonial product. Taiwan's Bureau of Cultural Heritage, Ministry of Culture, evaluates this mine as representative of the historic development of the Taiwan mining industry.[1] This perspective, however, leads to homogeneous

---

[1] https://twh.boch.gov.tw/taiwan/summary.aspx?id=6&lang=zh_tw#read, 3rd Feb 2019.

S. Hatano (✉)
University of the Ryukyus, Nishihara, Japan
e-mail: sohatano@tm.u-ryukyu.ac.jp

© Springer Nature Singapore Pte Ltd. 2020
A. Ginoza (ed.), *The Challenges of Island Studies*,
https://doi.org/10.1007/978-981-15-6288-4_5

treatment of the cultural heritage (regardless of the uniqueness of its background) and marginalization of its social, political, and cultural aspects. In other words, the Cultural Heritage Bureau's evaluation of the Jinguashi Mines disregards the historical reality of the region. Landscape restored based on such evaluation embodies the normativity of the state and risks downplaying of the multiplicity and plurality of values inherent to the region. For example, in the early days of mine development from the end of the 19th century to the 20th century, there were many contests and conflicts over land use and land ownership among the original inhabitants, mine owners and the Governor-General of Taiwan, as this paper will analyze. The manner in which the Japanese mining company seized land for its mining operation was contested by the subordinated Taiwanese people. If the industrial heritage from the colonial period veils and disregards such contestation and subordination (which was often the case in previous studies on the management of heritage from Taiwan's colonial period), it will lose historicity based on the social and political elements exclusive to colonial states; only the material environment will be maintained, and the value for the dwellers will be left behind.

This paper examines the interface between relics and heritage/landscape as a dissonant one. First, it will discuss the concept of landscape as an arena for contestation. Second, it will introduce the heritization of cultural landscapes in Taiwan and the present condition of Jinguashi Mines. Third, it will analyze the historical dissonance in Jinguashi Mines, especially the contested and subordinated past of landuse, through the analysis of cadastral maps and land registers produced during the period of Japanese rule. Fourth, it will consider the contemporary meaning of dissonance generated by heritization of Jinguashi, and also of the deviation from normative conditions. Fifth, it will discuss the interface between heritage and community development that maintains historicity and authenticity of the mining landscape.

Landscape as an Arena for Contestation

The interpretation of landscape will differ greatly depending on whether its formation and change are viewed as "consumption" or as "production" of the landscape elements by the background events. While research conducted from the former point of view[2] "might effectively reveal landscape's role as the upholder of certain ideological values, it would have little to say about the processes through which the landscape was made and produced" (Wylie 2007: 102). More specifically, it focuses on the political or social consumption of the existing landscape (or landscape that existed historically), such as how the landscape represented the ruler's authority, or how the dwellers of the landscape was acclimatized to the spatial strategy of the ruler. For this reason, Mitchell (1994: 7–39) criticizes the theories of landscape textuality as viewing the landscape only as an outcome and reflection of cultural values, thereby neglecting the process of landscape production. Viewing the landscape as something

---

[2] Such as Duncan (1990) and Cosgrove (1984).

that is 'already complete' is effective in elucidating how landscape represents and sometimes defends a certain concept (such as that of a politician) but fails to discuss the process through which landscapes are created and produced.

On the other hand, Ingold (1993) maintains that in the material dimension of landscape there exists a relationship in which elements of the past, present and future regulate each other, and that mutual regularity is what forms the characteristic of the landscape. For those who place emphasis on the productive aspect of the landscape, the key to elucidating the quality of the landscape is in the mutual regularity of elements formed at different times based on interpenetration of past, present and future. This assumes that interpenetration of time is incessant: landscape, a priori, is characterized by constant and nearly endless change. Studies based on this perspective focus on the ongoing social and economic relations in the production of landscape (Wylie 2007, 106), where landscapes are viewed as being constantly in production; that is open to change, alteration and contestation (Wylie 2007: 106).

Schein (2003: 202), focusing on the normative dimensions of landscape, criticizes that the "empiricist scientific tradition...that suggests that landscape is primarily the result of human activity...leaves the landscape itself out of social and cultural processes," and presents how landscape cannot exist apart from the day-to-day practices of the people in society. Furthermore, he maintains that landscape shall be viewed not as a result of various human activities, but rather as an "object" and a conceptual composition of the world, to be regarded as an important part of social and cultural processes. In his view, landscape assists in the acclimatization, embodiment, or normatization of social relations.

As these studies indicate, a focus on the productive aspect of landscape gives rise to opportunities for contestation which are not offered when the landscape is viewed as an "outcome". Study of the process in which social and cultural norms are embodied in the landscape affords a critical viewpoint on landscape.

Landscape is always open to change, alteration and contestation. Gramsci suggested that "a particular hegemonic regime was not a permanent order of things, but had to win consent to a negotiated ideological settlement with subordinated groups" (Brooker 2008: 213). Landscape studies based on the production theory focus on actions such as struggle, conflict, collision, conspiracy and negotiation which embody the relationship between the environment and the dwellers living in the landscape, as well as the relationship among the actors. Zukin (1991) remarked that themes of power, coercion, and collective resistance shape the landscape as a social microcosm; Mitchell (1996: 95) maintained that landscapes "operate to naturalize an unjust world"; and Harner (2001; 662) said landscape was "part of the hegemonic culture".

As evident in these discussions, the landscape is both an actual state and a representation. In other words, it is a form and a symbol. Landscape is an image as well as a lived place (Harner 2001: 663). At the same time, the symbolic aspect of landscape confronts the reality of the material world in a specific location. If conflicting social groups contest the spatial form and the symbolic meaning attached to it, social space and physical place will be restructured through hegemonic processes.

Taiwan's Experience in Creating Cultural Landscape as a Heritage—Normatization of the Value of the Landscape as a Cultural Heritage

Heritization of Cultural Landscape

In such areas, the hegemonic process is characterized by a phenomenon referred to in modern society as "heritization". Heritization is defined as the process and behavior in which an item is institutionally (or sometimes, non-institutionally) designated and registered by a group of people or an organization (especially a nation or municipality) that decides the said item to be important based on a certain set of values. In such process or behavior, the subject is sorted and selected based on a certain historical viewpoint (interpretation). While this is considered important from the standpoint of passing on historical items to the next generation, Hewison (1987) criticizes that heritage can be anti-democratic, conceal social and spatial inequalities, and alter or destroy buildings and landscapes that should be protected.

The point made by Hewison is extremely important. Landscape consists not only of diverse physical elements such as agricultural fields, houses, utility poles, roads, forests and rivers, but also of social and cultural meanings; it is an ambivalent presence that has both materiality and social/cultural properties. By identifying landscape as a heritage and proceeding with its designation, registration, protection and use, "normative" landscape is established, and existing landscapes are normatized. When landscape is designated and registered as a cultural heritage, it is guaranteed to some degree that it will be carried over to the future. A specific historical viewpoint serves as a concept device through which the past is projected from the present to sustainably shape the future landscape. The heritization of landscape legally guarantees the sustainability of the said landscape, but at the same time, the value of the landscape is abridged to one that supports a particular historical viewpoint, resulting in normatization of the "better" landscape.

In 2004, Taiwan enacted a law to protect cultural landscapes as cultural heritage. Since, it has certified the value of numerous cultural landscapes.

The Cultural Heritage Preservation Act of Taiwan regards Taiwan's cultural landscapes as tangible cultural heritage. Cultural landscapes are defined as locations or environments formed through longtime interactions between human beings and the natural environments.[3] Cultural landscapes that hold high value from the standpoint of history, aesthetics, ethnology or anthropology are registered as cultural heritage (Table 1).[4]

Furthermore, the Enforcement Rules of the Cultural Heritage Preservation Act specify that landscape preservation shall focus on above-ground facilities formed by humans. It lists a wide range of items to be preserved, to include: places where people conducted day-to-day activities such as agriculture and fisheries; legendary

---

[3] https://law.moj.gov.tw/ENG/LawClass/LawAll.aspx?pcode=H017000, 1st Feb 2019.
[4] Refer to Hatano and Hirasawa (2015) on cultural landscapes in Taiwan.

Table 1 Legal provisions pertaining to cultural landscape in Taiwan

Title: Cultural Heritage Preservation Act	
Amended Date: 27/07/2016	
Article 3(7)	Cultural landscapes: Locations or environments formed through longtime interactions between human beings and the natural environments, which are of value from the point of view of history, aesthetics, ethnology, or anthropology
Title: Enforcement Rules of the Cultural Heritage Preservation Act	
Amended Date: 27/07/2017	
Article 6	The cultural landscapes referred to in Item 7 of Subparagraph 1 of Article 3 of the Act include comprehensive landscapes or above-ground facilities formed by humans after extensive use of natural resources, for example, places of legends and myths, historical and cultural routes, religious landscapes, historic gardens, agricultural, forestry, fishery, and husbandry landscapes, industrial landscapes, transportation landscapes, water facilities, military facilities, and other places

places; and industrial, water, transportation and military facilities associated with more modern forms of livelihood (Table 1). By specifying that physical facilities are subject to protection and giving typical examples of landscape, the government is dictating that the focus is on the visible aspects of landscape and prescribing the normative landscape of the nation (region). This is trivialization of the concept and definition of landscape as discussed in the preceding chapter, and leads to formation of norms by the state (region).

Heritization thus defines value, and in so doing forms or reforms the normative landscape. This is certainly an important activity for the country. But as discussed above, landscape can be an arena for contestation; by abridging its value to something limited leads to the denial of all other values (i.e., official denial of the opportunity to contest). It must be recognized and accepted that landscape may simultaneously have multiple values.

Museum Establishment in Jinguashi

The Gold Museum in Jinguashi Mines was planned since 2002 and was formally opened on November 4, 2004. The aims and objectives of the Museum were as follows:

(A) To preserve and reproduce the history and culture of mining;
(B) To offer a natural venue for environmental education;
(C) To promote gold art and metal crafts to establish a creative industry; and
(D) To serve as a community eco-museum.[5]

To date, the Museum has been working with researchers to map out the scope of the cultural landscape (Fig. 1).

---

[5]https://www.gep-en.ntpc.gov.tw/xmdoc/cont?xsmsid=0G274571752019632449, 3rd Feb 2019.

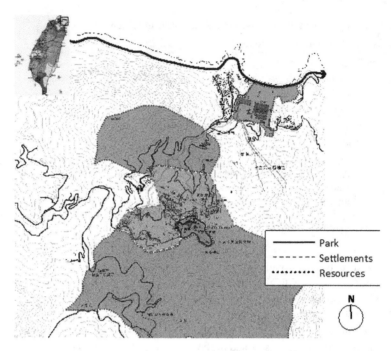

Fig. 1 Map of Jinguashi (*Source* Gold Museum, New Taipei City Government)

Despite the fact that one of the objectives of establishing the Museum is to preserve cultural properties, those in the subject area are still vulnerable to loss or deterioration. Although some of cultural properties such as the Crown Prince Chalet and Gold Shrine are registered by the local government, a number of them have been repaired by the owner without notification to the Cultural Affairs Department of the City Government. There are vulnerable conditions that the Park does not have control over, such as decaying houses (Figs. 2 and 3), mountain fire in 2010, and alterations to the former site of a hospital constructed in the early 1910s.

Jinguashi Mines as 'Dissonant Heritage'

As mentioned above, Taiwan Bureau of Cultural Heritage views Jinguashi as representative of 'the historic development of Taiwan's mining industry'. But what does 'historic development of the mining industry' mean? In contrast to colonies controlled by European countries, industrialization in Japanese-occupied colonies was predicated on advancement of Japanese companies into the industrialized colonies. Japanese people from political, economic and cultural fields entered the colonies as the main players of colonial development (Kobayasi 1993). In the process, a contact zone was formed, where Japanese people of different roles encountered the people

Fig. 2 Decaying Japanese style house (Photo taken on Sept. 6, 2012)

Fig. 3 Deteriorating Japanese style house (Photo taken on Sept. 6, 2012)

of Taiwan. Today, cultural heritage is not only a representation of former colonial power, but also a resource for tourism development. Therefore, the historical contact zone is closely tied to the protection and use of the cultural heritage; i.e., protection and utilization of the cultural heritage cannot be discussed without giving due consideration to the contact zone of the past. Pratt (2008: 8) defines the term 'contact zone' as the space of imperial encounters: a space in which people who are geographically and historically separated come into contact with each other and establish ongoing relations, usually involving conditions of coercion, radical inequality, and intractable conflict. From this perspective, Jinguashi Mines have different meanings to different people historically as well as presently: for the current government organization, it is a representation of industrial development; for the local residents it is a gold mine developed under Japanese rule; for some Japanese tourists the remains of Japanese style houses and shrine offer a sense of nostalgia; and for the Japanese mining company at the time it embodied the physical inequality between the Japanese and Taiwanese people, as seen in the conditions of landownership and landuse.

Colonialism was concerned with power, domination, and control. According to Palmer (1994), a dependence on tourism like the Bahamas serves to reinforce the historically implanted identity based on the artifacts of colonial occupation, so the tourism industry is in danger of perpetuating colonialism through the images. But this is not the case in Jinguashi where the colonial image has been diffused by application of purple paint on the outer walls of the Japanese style building known as the Crown Prince Chalet, as well as reuse of buildings from mixed time periods (including those built in the postwar era) by the Gold Museum.

But there is potential for dissonance in the heritage of post-colonial societies. According to Tunbridge and Ashworth (1996), a sense of dissonance is an intrinsic quality of heritage. Tunbridge and Ashworth (1996) argue that "all heritage is someone's heritage and therefore logically not someone else's". Dissonance has been characterized as the fundamental problem and opportunity for development of the tourist-historic city (Ashworth and Tunbridge 1990). Bruce and Creighton (2006) maintain that while heritage management of places having dissonance can stimulate economies, conserve built heritage, and strengthen local identities, it runs the risk of alienating host communities. Therefore, it is important to address the question, "who's heritage", and "conserving for what purpose". Discussions and negotiations for making decisions about conservation and re-use of buildings, institutions, relics of mining activities, and assembly landscape, must include the question of whether they should be seen as symbols of the colonial process and its consequences. In Jinguashi Mines, a sense of exclusion has created a psychological boundary between the Park and local residential area in addition to the physical boundary delineated by the Museum. At the same time, landownership is one of biggest question for local communities in identifying their own life and history. The current question of landownership is a historically generated one.

Based on this concept of dissonance, one must bear in mind that in the interpretation of heritage, historical objects and landscapes have different meanings and narratives for different groups. The following chapter examines this in further detail.

Dissonance in Landscape: Place for Multiple and Plural Values

Historical Dissonance

When Taiwan was ceded from Qing China to Japan as a result of the Sino-Japanese War, the Government of Japan actively promoted mine development through the Governor-General of Taiwan. In September 1896, the Governor-General of Taiwan recruited for Japanese mining companies to develop Mt. Jinguashi and Mr. Jiufen in the northeastern part of Taiwan.

The Tanaka Office led by Chobei Tanaka (1858–1924), who operated the Kamaishi Mine in Iwate Prefecture, Japan, won the mining rights to Mt. Jinguashi. The Tanaka Office immediately sent engineers and mine workers to the site and began mining near the summit of Mt. Jinguashi. Arefinery was built on a plateau below the summit. By about 1907, Jinguashi had turned into a modern mining town with multiple mining pits and galleries, refineries, residences and medical, educational and religious facilities for the Japanese engineers and mine workers (Hatano 2015a).

The mining company managed by The Tanaka Office expanded the mining territory to Suinantong, an area facing the sea in the northern part of Jinguashi. Before expanding the territory for its activities, the company in 1906 developed a plan to build a hydroelectric power plant and a watercourse on the farmlands and woods owned by Taiwanese farmers (see Fig. 4). To use the land, the mining company entered into contracts with the landowners agreeing to pay compensation in exchange for free use of the land by the company. Analysis of the historical materials pertaining to the contract indicates an unfair relationship between the company and landowners: (1) the farmers were able to use the remaining water for farming, but had to obtain permission from the company; (2) the company was allowed to alter the terrain if needed, for no additional compensation. The contract made it impossible for landowners to use their own land for farming without negotiating with the mining company. Landowners in Suinantong were forced to sell their lands to the company in the process of mining territory development, which followed maps drawn using cadastral maps and land registers (see Figs. 5, 6, 7 and 8). The maps indicate that the plots owned by the Taiwanese were sold to the mining company after 1916, and the company gradually enlarged their territory through the 1930s (also see Table 2).

The process of industrialization not only involved the development of mines, but also generated historical 'dissonance'. When the mines were designated as a heritage site and converted into a tourism spot, the colonial dissonance was overlooked. Heritage tourism should be concerned not only with preserving the remains of the past for visitor enjoyment, but also about contemporary struggles for power (Henderson 2001).

Fig. 4 Planning map for hydroelectric power plant and watercourse (*Source* Taiwan Historica)

Present Dissonance: Emergence of Multiple Values Accompanied by Dissonance

Who Is the Museum for?

In the postwar era, management of Jinguashi Mine was taken over by the Taiwan Metal Mining Corporation established by the Taiwanese Government in 1955.

Fig. 5 .

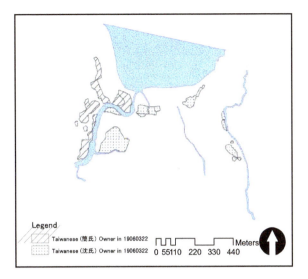

Fig. 6 .

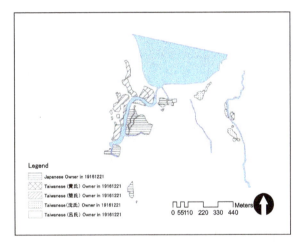

However, its debts grew, and the mine was closed in 1987. Today, the land of the former Jinguashi Mines is owned by Taiwan Power Company (Taipower) and Taiwan Sugar Corporation (Taisugar) who settled the debts. There is very little publicly-owned or privately-owned land.

After the closure of Jinguashi Mines, the local residents developed a plan to build a Gold Museum in an effort to vitalize their town. At the time, the residents recognized Jinguashi Mines as their own heritage (Hatano 2015b).

However, the said plan was absorbed into the museum construction plan of Taipei County (today's New Taipei City). Taipei County led the negotiation on land use with the landowners, and opened the Taipei County Gold Museum in 2004.

Fig. 7 .

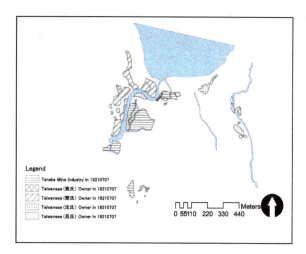

Fig. 8 .

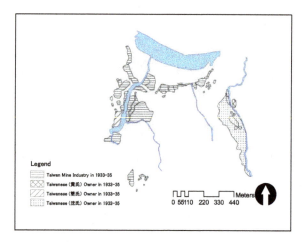

But the museum built by the Taipei County government did not allow for residents to participate in its operation. The residents, who should be responsible for running the museum, were deprived by the government of the initiative to establish the museum, and completely cast out of the museum itself.

Figure 9 shows stores built at Jinguashi. The shutters are located on the museum boundary; the road is inside the museum premises but the stores are outside. When the museum opened, some of the local residents built stores along the museum boundary with the entrance facing the museum, as an expression of spatial contestation against the spatial exclusion of local residents by the government authorities (Hatano 2019).

The local residents were also required to pay a high fee for land use, which became a source of friction between them and Taipower and Taisugar. The friction caused many residents to become indifferent about their town, resulting in abandonment of many houses which turned into ruins (also see Figs. 2 and 3).

Table 2 Transformation of landownership in Suinantong

	Shen (沈)		Jian (簡)		Huang (黃)		Lu (呂)		Mining Co.		National treasury		Total number	Total area (m^2)
	Number of property	Area (m^2)	Number of property	Area (m^2)	Number of property	Area (m^2)	Number of property	Area (m^2)	Number of property	Area (m^2)	Number of property	Area (m^2)		
22nd Mar 1906	16	32521	18	31789.7	0	0	0	0	0	0	0	0	34	64310.7
22nd Jan 1910	16	32521	18	31789.7	0	0	0	0	0	0	0	0	34	64310.7
21st Dec 1916	10	12333.3	20	32279.3	1	1290.7	1	244	14	37768.2	0	0	46	83915.5
7th Jul 1921	10	12333.3	11	32279.3	1	1290.7	1	244	19	38885.6	1	56.3	43	85595.9
1933–35	13	43697.3	3	5696.5	1	1290.7	0	0	37	75843.4	0	0	54	126527.9
14th Oct 1939	7	42648.7	3	5696.5	1	1290.7	6	1048.7	37	75843.4	0	0	54	126528

Fig. 9 Stores constructed by the residents (Photo taken on Sept. 6, 2012)

One of the exhibition facilities of the Gold Museum is an architecture called the Crown Prince Chalet. When the Crown Prince of Japan traveled to Taiwan in 1923, Jinguashi Mines were one of the places he planned to visit. The Chalet was built by the mining company as a resting facility for the Crown Prince (Fig. 10). It is a typical

Fig. 10 Wall painted in purple color (Photo taken on Sept. 6, 2012)

Fig. 11 Torii gate with depressions filled with new concrete (white parts) (Photo taken on Sept. 6, 2012)

example of Japanese architecture in Taiwan, and is designated as a cultural heritage of New Taipei City (Fig. 11).

In 2012, purple-color paint was applied to the entire outer wall of the Crown Prince Chalet by its owner, Taipower. Its authenticity as a cultural asset became questionable since the building was originally unpainted. The Museum immediately complained to Taipower through the New Taipei City government,[6] as it is illegal to alter a cultural heritage designated by the Cultural Heritage Preservation Act without permission. The New Taipei City government asked Taipower to restore the building to its original condition, but this only resulted in a slight dilution of the purple color (Fig. 10).

The Gold Shrine that remains within the premises of Gold Museum was built in the 1930s. The only parts of the original shrine that still remain are the foundation platform of the main hall, the pillars of the prayer hall, and two Torii gates on the approach road, which are designated as cultural heritage of New Taipei City. In 2012, the owner Taipower used cement-like material to fill the depressions in the prayer hall pillars and the Torii gates. The color is notably different from the original material, and looks out of place (Fig. 11). Again, the Museum complained to Taipower through the New Taipei City government, on the basis that it violated the prohibitions stipulated in the Cultural Heritage Preservation Act. However, the structure remained as is.[7]

---

[6]Based on interview with a curator of the Museum held on September 7, 2012.
[7]Ibid.

The Vivid Qitang Old Street

Nearby the Gold Museum is an old street called Qitang Old Street. The street was formed by houses built by the Taiwanese people who were engaged in the mining business of the Tanaka Office. In contrast to the Japanese residential area managed by the mining company, the Qitang Old Street area was reportedly a bustling town lined with various shops such as general goods stores, barber shops, and restaurants. The street exhibits characteristics of the Jinguashi Mines and has high historical value as cultural heritage. However, it has not been designated as a cultural heritage because it is a place where people currently live.

In the center of this town is a long stairway. In July 2017, a resident painted the railings of the stairs in colorful colors of red, blue, yellow, and green (Fig. 12). The resident apparently wanted to make the street more attractive and beautiful, but this action was criticized on the Internet as "destruction of old townscape", "sad", and "ugly" (呉淑君 2018). The area is outside the premises of the Gold Museum and is not designated a cultural heritage, so the Museum had no means to file a complaint. Today, tourists visit the stairway to take photos to post on SNS, which has helped turn the townscape into a tourism resource.

Fig. 12 Stairs with railing painted in bright colors (Qitang Old Street)(Resource: Gold Museum, New Taipei City Government)

Conciliation and Dialogue in the Landscape

As discussed above, the landscape of Jinguashi Mines was contested as it was shaped through history, and continues to be contested today which cannot not be disregarded. There is need to go back to the question of, "whose heritage". Dialogue among different groups can help discover the different interpretations of heritage and assist in building place-based authenticity.

According to Bruner (1994), authenticity can be categorized into four types: 'verisimilitude', 'genuineness', 'originality', and 'authority'. 'Verisimilitude' and 'genuineness' imply the item is not in the same condition as the original object, but has the atmosphere of the 19xx's or has been rebuilt to nearly the same as the original. 'Originality' means that buildings and objects are original in history. In this sense, 'originality' is equivalent to "authenticity in heritage" as described in the Operational Guidelines for the Implementation of the World Heritage Convention, such as the authenticity of design, function, material, design, and so on. 'Authority' is heritage authorized by the government or other organizations even if the said heritage is merely 'genuine'. In this sense, 'authenticity' is not only intrinsic to the physical objects and landscapes themselves but is also constructed socially through the interpretations of different groups and the negotiations between them. According to Waitt (2000), authenticity is no longer treated as a fixed concept, but is a theme of negotiation and continuous reinterpretation.

In Jinguashi Mines, the Museum's revitalization project focused on building façades and stairways along the Qitang Old Street that the Taiwanese people constructed during the period of Japanese occupation (Fig. 13). The town was dressed to provide a 'look' for the visitors. Although the project did not involve historical research on architecture or artificial environment, the residents were engaged in the discussions to produce a guideline for redesign of architecture. While this method was not conducive to conserving the original heritage, the residents and the Museum discussed to find authenticity in the façade through the use of their own historical knowledge, as White (2014) says, 'the practical past', and reconstructed a town that felt like the place where they played during childhood. This can be seen as a construction of their own authenticity—place-based authenticity, or more generally, constructive authenticity (Wang 1999) (Fig. 14).

But there is still a lack of system for interpretation and dialogue between the different groups. A system of dialogue must be developed to promote expression of historical dissonance and to mitigate the dissonant behavior still observed today.

Based on the research of Jinguashi Mines, the mining landscape can be interpreted from three different perspectives, namely: the physical environment, tourism attractions, and themes. The physical environment, which changes from the past to the present and in the future, is intrinsically tied to the mining history. Some buildings are authorized as heritage by the government, while others are just remains or relics of the past. Some of buildings have been converted into museum institutions as tourism attractions. The Museum interprets and introduces relics of mining activities but is not responsible for conserving them. Jinguashi Mines offer encounters with

Fig. 13 Reformed Wall at Qitang Old Street

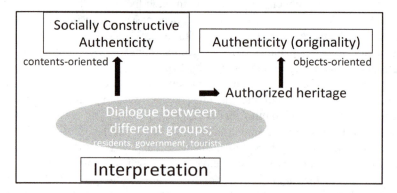

Fig. 14 Socially-constructed authenticity

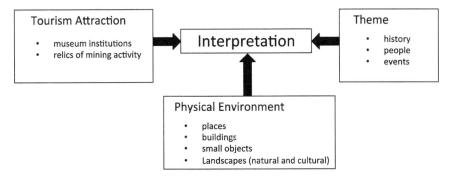

Fig. 15 Interpretation of landscape in Jinguashi

different themes such as history, people, and events, which are subject to different interpretations by the visitors and dwellers (Fig. 15).

Systematic heritization tends to focus on the tangible aspects of the mines, which offer an easy-to-understand account of the historical development of the mining activities. Community identities and evaluation of the historical properties are often related to the physical remains (Bruce and Creighton 2006). But the intangible aspects of the heritage also should be identified through historical research and narratives of the residents.

Conclusion

Jinguashi Mines should be recreated as a mixed landscape, combining the landscape created by colonial rulers and the landscape that the locals experienced as a lived place. Today, the changes in land use that were forced upon by Japanese rule are no longer visible; yet they should be presented as 'heritage as knowledge' (Graham 2002).

This paper suggests the following conceptual framework for interface between conserving heritage and developing tourism in Jinguashi Mines:

1. Post-colonial societies working with cultural heritage should ask the question, 'whose heritage' and 'conserving for what purpose'. Discussions and negotiations for decision-making concerning the conservation and re-use of buildings, institutions, relics of mining activities, and assembly landscape must include the question of whether they should be viewed as symbols of the colonial process and its consequences.
2. Management of the heritage territory requires the perspectives of both the tourists and residents. The management of Jinguashi mining landscape and tourism, from the perspective of tourists, can be divided into three steps. First, in the early stage of tourism development, the expectation of the tourists is to see the well-known history of the mines (industrial development during early modern period). In

the second step, the expectation of the tourists expands towards understanding diversification and complexity of colonialism. In the third step, the community experiencing the second step acquires the true historicity of the mining landscape; i.e., the community will not only 'live in', but also 'live with' the historicity of colonialism.
3. From the community viewpoint, heritage management can provide an opportunity for dialogue among residents, professionals of heritage conservation and tourism, and the administrative agency. In the dialogue, due attention shall be given to the interface between industrial development and colonial statements.

Landscapes, no matter how they are normatized as cultural heritage, cannot exist unrelated to the daily lives of the people. In other words, landscapes do not exist external to peoples' activities and consciousness; they emerge from the process of interpenetration between the physical environment and the daily lives of the people. In a place where a human group politically and socially oppresses another human group, the physical environment is restructured to embody such oppression. Thus, the landscape becomes a representation of the political condition. At times, landscape effectively serves to promote and sustain the oppression; at the same time, such representation may elicit spatial criticism and/or spatial contestation.

As this paper shows, heritization (=normatization) of landscape based on a historical view that is different from the region's historical or current reality will give rise to contestation. The spatial contestation is based on the "practical past" as described by White (2014), and presents the plurality of the past. In other words, the landscape is inherently capable of serving as a mediator of social relations such as cooperation, collaboration, competition and rivalry among the multiple and potentially dissonant narratives of the past and present. Such function of landscape should be fully exploited, particularly in island areas.

References

吳淑君 (2018) 「祈堂老街被漆上繽紛顏色彩網友嘆: 金瓜石要毀了」『聯合新聞網』. https://udn.com/news/story/7266/3390266, written in Chinese.
Ashworth, G. J., & Tunbridge, J. E. (1990). *The tourist-historic city: Retrospect and prospect of managing the heritage city*. London: Belhaven.
Brooker, P. (2008). Hegemony. In *A glossary of literary and cultural theory* (pp. 212–213) (K. Arimoto, & T. Motohashi, Trans.). Shinyousha.
Bruce, D., & Creighton, O. (2006). Contested identities: The dissonant heritage of European town walls and walled towns. *International Journal of Heritage Studies, 12*(3), 234–254.
Bruner, E. (1994). Abraham Lincoln as authentic reproduction: A critic of postmodernism. *American Anthropologist, 96*(2), 397–415.
Cosgrove, D. E. (1984). *Social formation and symbolic landscape*. London: Croom Helm.
Duncan, J. S. (1990). *The city as text: The politics of landscape interpretation in the Kandyan Kingdom*. Cambridge: Cambridge University Press.
Graham, B. (2002). Heritage as knowledge: Capital or culture? In *Urban studies* (Vol. 39, pp. 1003–1017).

Harner, J. (2001). Place identity and copper mining in Sonora, Mexico. *Annals of Association of American Geographers, 91*(4), 660–680.
Hatano, S. (2015a). The actual arrangement of spaces and facilities in the mines of Ruifang and Jinguashi During the Meiji 30s (1897–1906). *Gold Museum Journal, 3*, 50–70, written in Chinese.
Hatano, S. (2015b). Taiwan niokeru Bunka Keikan no Isanka (Heritization of cultural landscapes in Taiwan). In M. Segawa (Ed.), *World heritages and cultural resources in Northeast Asia* (pp. 77–86). Sendai: Center for Northeast Asian Studies Tohoku University, written in Japanese.
Hatano, S. (2019). Bunka Isan ha Dare no Mono Ka (Who does own cultural heritage?).
Hatano, S., & Hirasawa, T. (2015). Taiwan no Bunka-Keikannimiru Kukan, Hou, Syakai (Space, law and society in cultural landscape of Taiwan), in Isekigaku Kenkyu. *Journal of the Japanese Society for Cultural Heritage, 12*, 114–119, written in Japanese.
Henderson, J. C. (2001). Conserving colonial heritage: Raffles hotel in Singapore. *International Journal of Heritage Studies, 7*(1), 7–24.
Hewison, R. (1987). *The heritage industry*. York: Methuen Publishing.
Ingold, T. (1993). The temporality of the landscape. *World Archaeology, 25*(2), 152–174.
Kobayasi, H., et al. (1993). *Iwanami-Kouza Kindai Nihon to Syokuminchi 3 Syokuminchi to Sangyouka (Colonization and industrialization)*. Tokyo: Iwanami Syoten, written in Japanese.
Massey, D. (2005). *For space*. London: Sage Publications.
Mitchell, D. (1994). Landscape and surplus value: The making of the ordinary in Brentwood, California. *Environment and Planning D: Society and Space, 12*, 7–30.
Mitchell, D. (1996). Sticks and stones: The work of landscape. *Professional Geographer, 48*(1), 94–96.
Palmer, C. A. (1994). Tourism and colonialism: The experience of the Bahamas. *Annals of Tourism Research, 21*(4), 792–811.
Pratt, M. L. (2008). *Imperial eyes: Travel writing and transculturation*. London: Routledge.
Schein, R. H. (2003). Normative dimensions of landscape. In C. Wilson, & P. Groth (Eds.), *Everyday America: Cultural landscape studies after J. B. Jackson* (pp. 199–218). California: University of California Press.
Tunbridge, J. E., & Ashworth, G. J. (1996). *Dissonant heritage: The management of the past as a resource in conflict*. Chichester: Wileys.
Waitt, G. (2000). Consuming heritage: Perceived historical authenticity. *Annals of Tourism Research, 27*(4), 835–862.
Wang, N. (1999). Rethinking authenticity in tourism experience. *Annals of Tourism Research, 26*(2), 349–370.
White, H. (2014). *The practical past*. Evanston: Northwestern University Press.
Wylie, J. (2007). *Landscape*. London: Routledge.
Zukin, S. (1991). *Landscapes of power: From Detroit to Disney World*. California: University of California Press.

The Possibilities of Phylogenetic Tree Studies in Ryukyuan Languages Research

Shigehisa Karimata

The Position of the Ryukyuan Languages

The Ryukyu Archipelago consists of four island groups, namely the Amami Islands in Kagoshima Prefecture and the Okinawa, Miyako, and Yaeyama Islands in Okinawa Prefecture. The languages that have been traditionally spoken in the Ryukyu archipelago are called the Ryukyuan languages. Based on previous research findings, it is widely known that Ryukyuan diverged from proto-Japanese-Ryukyuan. It is the only language proven to have genealogical relationship with Japanese.

Ryukyuan is a minority language, with the population of the Ryukyu Archipelago being only 1% of Japan's overall population. Nevertheless, Ryukuan is recognized as having a contrasting relationship with Japanese because Ryukyu was an independent country with its own history for approximately 450 years until it was integrated into Japan in 1879, during which time Ryukyuan underwent unique development without significant influence from Japanese and became quite different from the Japanese language and its dialects.

The Ryukyu Archipelago's total land area is less than 1% of Japan's overall land area; yet its 47 inhabited islands of various sizes are scattered over a vast ocean area of approximately 900 km. When a map of the Ryukyu Archipelago is overlaid on a map of the main island of Japan placing Kikaijima on top of Sendai City, Amami Oshima corresponds to Yamagata Prefecture, Naha City on Okinawa Island overlaps with Nagano Prefecture, Miyako Island is in between Kyoto City and Osaka City, and

---

S. Karimata (✉)
Research Institute for Islands and Sustainability, University of the Ryukyus, Nishihara, Japan
e-mail: karimata@ll.u-ryukyu.ac.jp

Yonaguni Island (Japan's westernmost island) lies between Okayama and Hiroshima Prefectures.[1]

---

[1] When the maps of Japan and Europe are overlapped with each other, it you can see Japan is long north to south and has diverse climate and nature. Cape Soya in the northern tip of Hokkaido is Oslo in Norway and Nemuro Peninsula in eastern Hokkaido is Stockholm in Sweden. Yonaguni in Okinawa (westernmost island in Japan) is Gibraltar of southern Spain near Morocco of the African continent. Japan is a long island nation consisting of islands of various sizes spread out over a wide climate range from the subarctic and subtropical zones.

Research Goals of Ryukyuan Language Studies

Following the split from proto-Japanese-Ryukyuan (pJR), people who migrated south from Kyushu spread the language to the Ryukyu Archipelago, where the language underwent unique development in the respective regions.

There are vast differences among the various Ryukyuan languages. The dialects of Kikaijima in the northern edge of the Ryukyu Archipelago and Yonaguni at the southwest end are mutually unintelligible. Likewise, the languages are not mutually intelligible between Yonaguni and Ishigaki Island, Miyako and Okinawa Islands, or between Amami and Okinawa Islands. If standard Japanese was not spoken, people would need an interpreter to communicate with each other.

When did the people who spoke proto-Ryukyuan migrate from Kyushu to the Ryukyu Archipelago, and what was the scale or frequency of their migration? When did people migrate to the southernmost island of Hateruma or to the westernmost island of Yonaguni, and in what scale? How did the language spread within the Archipelago and develop in their respective areas? These questions still remain unanswered. A strong relationship with the Kyushu dialects is suggested; however, the linguistic characteristics that date pack to proto-Kyushu-Ryukyuan have not been clearly distinguished from the linguistic influence following the Satsuma invasion of 1609.

If languages not only diverge but also influence each other through contacts and mix in complex ways to characterize each dialect, then it is important to clarify how the contacts between various languages influence the subordinate dialects.

A major challenge for Ryukyuan language studies is to look at the Ryukyuan languages holistically utilizing the accumulation of research on Ryukyuan languages to date, and build a framework for systematic and comprehensive study of the Ryukyuan languages. The goal is to trace which subordinate languages diverged from which superior language, identify their characteristics, and shed light on their formative factors and processes.

Linguistic Differences Between the Northern and Southern Ryukyuan Languages

Ryukyuan is largely categorized into two groups: Northern Ryukyuan (NR) spoken in Amami and Okinawa Isalnds, and Southern Ryukyuan (SR) used in Miyako and Yaeyama Islands. The differences between the two groups span across all linguistic features, such as phoneme, grammar and vocabulary. Why and how these differences emerged was unknown.

Karimata (2018) compared the grammatical and vocabulary differences between NR and SR, and further examined what impact the Kyushu dialects had on the difference between the two Ryukyuan languages.

Karimata (2018) investigated the linguistic differences between Northern and Southern Ryukyuan focusing on grammatical features, which are thought to be more conservative than vocabulary. The main grammatical phenomena investigated were: (1) onbin, or euphonic changes observed in some inflected verbal forms like past forms (Table 1), (2) aspectual system and its constituting verbal forms (Table 2), (3) the inanimate negative existential and its grammatical functions, and (4) converb forms functioning as head of adpositive and conjunctive subordinate clauses.

(1-1) Kyushu dialects and Northern Ryukyuan are characterized by euphonic changes (nasalization, gemination, u- and i-vocalization) in some verbal inflected forms like past. There are no euphonic changes in Southern Ryukyuan.

(1-2) Southern Ryukyuan is the only language branch not showing any euphonic changes. As such, it retains an archaic feature of Japonic languages.

(1-3) The euphonic changes observed in Northern Ryukyuan are the same as those observed in Kyushu dialects.

(2-1) The aspectual system of Northern Ryukyuan is formally a ternary system (suru, shiyoru, shitoru) like Kyushu dialects, while that of Southern Ryukyuan is a binary system (suru, shiteiru).

(2-2) The Northern Ryukyuan verbal form cognate with "shitoru" expresses progressive and resultative aspect, contrary to Kyushu dialects but like Eastern Japanese and Southern Ryukyuan.

(2-3) The Northern Ryukyuan verbal form cognate with "shiyoru" expresses perfective aspect, like Eastern Japanese "suru".

Table 1 Euphonic changes in Southern and Northern Ryukyuan

Kyushu dialects	Northern Ryukyuan			Southern Ryukyuan	
	Omoro Soshi[a]	Amami	Okinawa	Miyako	Yaeyama
Yes	Yes	Yes	Yes	No	No

[a]Karimata (2006) discussed the past tense and converb cognate with *shite showing euphonic changes in the Okinawa dialect and Omoro Soshi

Table 2 Aspectual system (nom- "drink" and oki- "wake up")

Kyushu dialects		Naha		Miyako	Ishigaki	Eastern Japan	
nomuokiru	PRF	numun ukiin	PRF	num uki	numun ukin	nomu okiru	PRF
nomiyoru okiyoru	PROG						
nondoru okitoru	RES	nudoon ukitoon	IPF	numiuu ukiuu	numin ukeen	nondeiru okiteiru	IPF

PRF Perfective; *PROG* Progressive; *RES* Resultative; *IPF* Imperfective

(3-1) The Inanimate negative existential is expressed by an irregular verb in Ryukyuan, but by an adjective in Kyushu dialects (and in all other Japanese dialects).

(3-2) In Ryukyuan, the inanimate negative existential verb has, beside its original function, grammaticalized into a completive aspect auxiliary.

(3-3) The inanimate negative existential does not simply share formal characteristics with verbs, it behaves morpho-syntactically like any other verb.

(3-4) The inanimate negative existential is an important criterion to distinguish Ryukyuan from Japanese. There is, however, a negative auxiliary verb nap-2 in Eastern old Japanese.[2]

There are two non-cognate converbs functioning as head of adpositive and conjunctive subordinate clauses in Ryukyuan languages: one going back to *shite (cognate with modern Japanese te-converb) and one to *shiari.

(4-1) Among Northern Ryukyuan languages, dialects spoken in Amami islands and the Northern part of Okinawa have a converb cognate with *shite, showing the same euphonic changes as Kyushu dialects.

(4-2) Southern Ryukyuan and the Northern Ryukyuan dialects spoken in Iheya and Izena islands have a converb cognate with *shiari, but show no trace of a converb cognate with *shite.

(4-3) The dialects spoken in South and central Okinawa and the language represented by *Omoro Sōshi*, which was edited around 1500, have a converb cognate with *shite showing euphonic changes but also a converb cognate with *shiari.

Based on the fact that there is a significant difference between Northern Ryukyuan (NR) and Southern Ryukyuan (SR) languages, and that NR and Kyushu dialects have common characteristic features, it was concluded that the language spoken in Japan before the Nara period spread to Southern Ryukyu, while the language spoken after the Heian period spread to Northern Ryukyu.

Karimata (2019) examined vocabulary differences between NR and SR using the words "rice", "rice plant", "cooked rice" and "sickle". As shown in Table 3, SR does not distinguish rice plant from rice: a cognate of "mai" is used to mean both. In NR, rice plant (*ini) and rice (*kome) are lexically distinguished.

In the Northern Ryukyuan dialects spoken in the Northern area of Okinawa (Kunigami, Ogimi, Higashi, Nakijin[3] and Motobu), rice and rice plant are lexically distinguished, rice being referred to with a cognate of *kome, while rice plant is referred to with a cognate of *mai.[4] In Yoron island, which is situated just North of Okinawa's main island, the dialect spoken in Gusuku does not distinguish between

---

[2]For example, [tati-wakareini-siyopiywori se-roniapa-napuyo] "I haven't met you since the night when you took your leave" (Manyoshu 3375). Although nap- exhibits verbal-like behavior, it should be noted that, whereas in Ryukyuan the negative existential is a free verb, it is a suffix in Eastern old Japanese. Both forms may constitute an example of convergence.

[3]In the NakijinYonamine dialect, *me (rice crop) is a component of compound words such as me:hai (rice reapking) and me:gadʒimi (rice straw).

[4]Nago City History Compilation Committee 2006. See especially pp. 332–333 and pp. 492–493.

Table 3 Words related to rice and sickle

	Rice	Rice plant	Cooked rice	Sickle
Yamatobama, Amami Is.	kumï	ʔniː	ʔobaN	kama
Boma, Tokunoshima Is.	kumï	ʔiniː		kama
Gusuku, Yoron Is.	mai	mai	mai	hama
Nakijin Yonamine, Okinawa Is.	humiː	meː	meː	hamaː /iraːna
Naha Shuri, Okinawa Is.	kumi	ʔNni	meː /ʔubuN	ʔirana
Hirara Shimozato, Miyako Is.	maz	maz	maz	zzara
Ishigaki, Yaeyama Is.	maɿ	maɿ	maɿ/NboN	gagɿ
Sonai, Yonaguni Is.	mai	mai	mai	irara

rice and rice plant both meanings are referred to with a single word, mai. Sickle (kama), a word related to rice and rice plant, was also examined, and kama was found in NR and irara was found in SR. Cognate of irara was found in the Northern part of Okinawa.

Based on the above, Karimata (2019) proposed a hypothesis that there were two waves of migration from Kyushu that caused the linguistic differences between Southern and Northern Ryukyuan languages.

(A) A group of people who spoke proto-Kyushu/Ryukyuan, a language characterized by a lack of euphonic changes, a converb going back to *shiari and a binary aspectual system, departed from South Kyushu and spread into Amami and Okinawa islands.
(B) Their language had the words *pise (reef) and *kanoku/kaneku (sandy place).
(C) These people cultivated *mai (rice), making no lexical distinction between rice and rice plant, and used *irara (sickle) to reap it.
(D) After the first movement Southward, they crossed the Kerama gap into Miyako and Yaeyama islands.
(E) Long after the first migration wave, a second group of people, who spoke a language characterized by euphonic changes, a converb going back to shite and a ternary aspectual system, migrated from Kyushu and constituted the second migration wave.
(F) These people cultivated *ine (rice plant) and used *kama (sickle) to reap *kome (rice). Their language exerted a profound influence on the languages spoken from Amami to Okinoerabu islands.
(G) In the areas from Yoron island to Northern Okinawa, the word for rice, the main crop and the crop used for tribute payments, was substituted for *kome, while *mai continued to be used for rice plant.
(H) In Central/South Okinawa, people cultivated *ine (rice plant) and reaped *kome (rice) with *irara (sickle). In these areas, a language emerged possessing both a shiari and a shite converb, and an aspectual system characterized by a mix of a ternary and a binary system.

Although Karimata (2018, 2019) examined the characteristic differences in grammar and vocabulary, the study examined only limited items. To verify Karimata's hypothesis, the number of items subject to examination needs to be increased, and studied from multiple aspects. Similarly, the characteristic features of phoneme, grammar and vocabularies must be examined to understand dialects in other subordinate dialects.

Linguistic Geography in Ryukyuan Language Studies

The Okinawa Language Research Center (OLRC), a private organization, prepared the "Basic Questionnaire (1979)" capable of investigating 1103 words as part of its plan to record and preserve the endangered Ryukyuan languages.

Subsequently, the Research Center added a plan to survey the dialects of all traditional settlements, and prepared the "All Settlements Questionnaire for Study of the Languages of the Ryukyu Archipelago (1983)" (hereafter referred to as "All Settlements"). The All Settlements Questionnaire extracted from the Basic Questionnaire "the parts of the community's dialect that constitute its structural basis, such as the phonemic system and basic or everyday vocabulary", and added verbs and adjectives to identify their general conjugation types. It includes 200 words, with multiple questions per word, for a total of 350 survey items (including grammatical form of conjugated words).

The Research Center developed a plan to survey 100 locations in the Ryukyu Archipelago using the Basic Questionnaire and 700 settlements using the All Settlements Questionnaire. The surveys were conducted from 1979 to 1992 using multiple external funds. Survey results were obtained from 628 locations for All Settlements; 99 locations for Part 1, 114 locations for Part 2, 82 locations for Survey No. 2, 61 locations for Survey No. 3, and 56 locations for Survey No. 4 for basic surveys.

One of the methods to gain a holistic view of a target local language is linguistic geography, which studies the variations and development of languages on spatial coordinate axes based on the geographical distribution of the language phenomena. Since languages have systematic structures, the survey items are selected based on the system and structure. The linguistic geography study by the Okinawa Language Research Center use systematic approaches in the selection of survey items, following the structurally comparative linguistic geography suggested by Nakasone (1937), who organically connected linguistic geography with comparative linguistics.

While there is much to gain from linguistic geography studies, the shortfall of a linguistic map is that it is two-dimensional, making it difficult to consolidate multiple items on a single map, or to show the relationship between two distantly-positioned local languages.

Ryukyuan Dialect Dictionary

The Ryukyu Archipelago and Japan have a long history of mutual interaction even after the split of Ryukyuan language from proto-Japanese-Ryukyuan. Assuming such historic interaction is reflected in the local languages, there should be no deep rupture between the Ryukyuan and Japanese languages. The boundary between the Ryukyuan and Japanese languages is not a single line; there are multiple lines drawn between Kyushu, Tokara Islands and the Ryukyu Archipelago as well as within the Ryukyu Archipelago. The boundary lines indicate the trails of people's southward migration and differences in the degree and number of linguistic contact after the split. The history of the formation and development of the Ryukyuan languages can be elucidated only by carefully peeling the multiple thin layers of such information.

Of the common features between the Ryukyuan and Kyushu dialects, some go back to the protolanguage while others are related to linguistic contact after the split. The level of influence of the Kyushu dialects on the Ryukyuan languages varies by the subordinate dialects: Amami dialect was affected greatly while Yonaguni dialect received little influence. Comparative studies of the basic vocabularies alone are not sufficient to elucidate the possibility of proto-Kyushu-Ryukyuan or the degree of influence of linguistic contacts.

Fortunately, there has been ten Ryukyuan dialect dictionaries published to date, each containing more than 10,000 words. There are two additional dictionaries that contain less than 10,000 words and one more that is not in public print, for a total of 13 Ryukyuan dialect dictionaries. Two Kagoshima dialect dictionaries have also been published. If these dictionaries can be combined in a database that can be searched by (1) dialect form, (2) standard language form, (3) semantics, and (4) key words, it will enable comparative historical linguistic studies of the different Ryukyuan dialects well as between the Ryukyuan and Southern Kyushu dialects.

1. *Amami Dialect Classification Dictionary, Vol. 1 and Vol. 2* (Nagata, Suyama and Fujii 1977 and 1989)
2. *Dictionary of the Amami Tatsugo Dialect* (Ishizaki, unpublished)
3. *Dictionary of the Yoron Dialect* (Chiyo and Takahashi 2004)
4. *Dictionary of the Izena Dialect* (Izena Village Board of Education 2004)
5. *Dictionary of the Okinawa Iejima Dialect* (Oshio 1999)
6. *Dictionary of the Okinawa Nakijin Dialect* (Nakasone 1983)
7. *The Okinawan Language Dictionary* (Compiled by National Language Research Institute 1963)
8. *Dictionary of the Miyako Irabu Dialect* (Toyohama 2013)
9. *Dictionary of the Ishigaki Dialect* (Miyagi 2003)
10. *Dictionary of the Taketomi Dialect* (Maeara 2010)
11. *The Yonaguni Language Dictionary* (Ikema 1998)
12. *Dictionary of the Tanegashima Dialect* (Uemura 2001)
13. *Dictionary of the Kagoshima Dialect* (Hashiguchi 2004)

Phylogenetic Tree Studies in Ryukyuan Languages

Based on the results of the Ryukyuan language research to date, a joint research[5] was initiated to explore the history of divergence and interaction among the subordinate Ryukyuan languages as well as the history of divergence from the proto-Japanese-Ryukyuan language and the interaction with Ryukyuan dialects. This was achieved by drawing a phylogenetic tree based on the "Basic Questionnaire" and "All Settlements Questionnaire" collected by the Okinawa Linguistic Research Center as well as the words listed on the 13 dialect dictionaries.

Of the 97 dialects selected from the entire Ryukyus, 72 randomly selected words were broken down into phoneme; each distinctive feature of phoneme was digitized (0 and 1) and connected to develop a prototype phylogenetic tree using Neighbor-net and Neighbor-Joining. For words with different radicals, separate sheets were created and integrated. The phylogenetic tree created by digitizing the distinctive features of phoneme was confirmed to be valid to a certain extent.

Phylogenetic trees constructed by Neighbor-net and Neighbor-Joining (72 words)

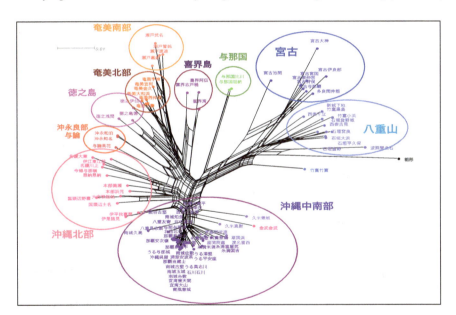

---

[5]Researchers include: Ryosuke Kimura (Univ. of the Ryukyus, Genomic Anthropology), Takeo Okazaki (Univ. of the Ryukyus, Mathematical Statistics), Nobuko Kibe (National Language Research Institute, Japanese), Akihiro Kaneda (Chiba University, Japanese), Hiroomi Tsumura (Doshisha University, Spatial Information Science), Michinori Shimoji (Kyushu University, Linguistics), Rihito Shirata (Shigakukan University, Linguistics), Yukiko Shimabukuro, Keiko Nakama, Nana Toyama, Gils, vander Lubbe, and Kaishi Yamagiwa.

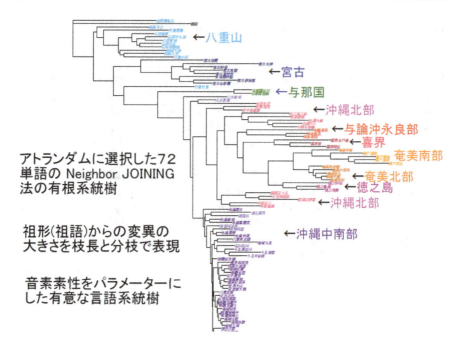

Based on the knowledge gained through phonological studies of the Ryukyuan languages and the results of linguistic geography studies to date, the current study focuses on confirming the validity of the phylogenetic tree using digitized phoneme features, as well as developing improved methods for constructing the phylogenetic tree. A phoneme is expressed as set of distinctive features; a word is expressed as a continuum of digitized sets of distinctive features. Phonologic changes are expressed as variation of the distinctive features. A phylogenetic tree constructed by breaking down phoneme into distinctive features is a tree of traits that shape the external forms of words. Very few words in the Ryukyuan language are composed of a single vowel; an overwhelming majority of Ryukyuan words are composed of multiple syllables made of consonant and vowel. By using digitized distinctive features of phoneme, a phylogenetic tree can be constructed using a single word.

A phylogenetic tree constructed by a single digitized word depicts the process of emergence of subordinate dialects from the protolanguage. The shape of a single-word phylogenetic tree shows the process of change from the protolanguage. A single-word phylogenetic tree is constructed based on a specific trait. A careful examination is required to understand what the single-word tree indicates. Some phonological changes occur regardless of sound environments, such as the difference in rhyme of the preceding or succeeding phoneme or the difference in position within the word (anlaut, inlaut or end), while other phonological change occurs under limited conditions and sound environments. Some phonological change, such as $^*o > u$, occurred in the entire Ryukyuan languages, while other phonological change, such as $^*p > h$, was observed in two distant areas. These changes are treated equally in the current study, but there is need to appropriately weight widespread changes versus rare changes, as well as to select words that are better suited for construction of the phylogenetic tree. Efforts must be made to avoid any arbitrariness.

Once the validity of the single-word phylogenetic tree based on distinctive features of the phoneme is confirmed, it can be used to study which elements are effective for phylogenetic trees that integrate multiple words. It should then become possible to produce phylogenetic trees from freely selected word combinations or by connecting multiple words that are substitutable.

A linguistic map can be drafted by plotting end points of the phylogenetic tree on a geographical map using GIS data. This map can be used to validate the adequacy of the phylogenetic tree based on linguistic geography research findings. The linguistic map drawn based on linguistic phylogenetic trees will introduce a three-dimensional perspective to linguistic geography, and will dynamically depict on coordinates of time and space the history of how the proto-Japanese-Ryukyuan and proto-Kyushu-Ryukuan languages branched off to form the Ryukyuan languages. In addition to the single-word linguistic map, a small system of language can be drawn on linguistic maps using specific phylogenetic trees constructed from multiple words. Such linguistic maps will be more comprehensive, structural, and systematic.

Dialect and Folk Culture

Variations within Southern Ryukyuan (SR) languages are relatively small. It can be categorized into the Western Shimajiri (WS) and Eastern Shimajiri (ES) dialects. WS has no gutturalized approximant, ?j and ?w, while ES has gutturalized approximant of ?j and ?w. The word "elder brother" is referred to as *aiya* in WS and *appi* or *afi* in ES dialect. The Southern Ryukyuan dialect (SR) is subdivided into East and West based on phonologic characteristics as well as lexical properties.

The WS dialect is spoken in today's Itoman City, which is equivalent to the area referred to in *Komesu-Omorono-on-soshi* in Vol. 20 of *Omoro Soshi*. The ES dialect is spoken present day Yaese Town and Nanjo City, which equates to the are as referred to in *Shimanaka Omoro-on-soshi* (Vol. 18) and *Chienen Sashiki Hanakusuku Omoro-on-soshi* (Vol. 19) of *Omoro Soshi*, respectively. Archeologist Susumu Asato (1990) used pottery clay classifications to categorize Shimajiri, the southern part of Okinawa Island, into three zones: Class B dominant areas, Class A and C dominant areas, and Class D dominant areas. Class A and C dominant areas overlap with the ES dialect area, while Class D dominant areas overlap with WS dialect area.

This indicates that the administrative divisions introduced in the Gusuku Period were carried over to the period of Omoro Soshi compilation, and to today's dialect zones. If the origin of WS and ES dialect zones dates back to the Gusuku Period, the dialect zones in other regions may likewise be traceable to the Gusuku Period; e.g., the dialect zones in northern Okinawa may corresponds to the administrative division called *magiri* which was established during the Ryukyu Kingdom era.

According to Nakasone (1937), the attributive form of verbs end in "nu" in areas north of Onna Village on Okinawa Island, whereas the areas south of Onna uses "ru" instead. In the areas where "nu" is used, the word *ashagi* refers to a religious structure built in the community's ceremony site; in areas where "ru" is used, the *ashagi* is an independent building inside a residential property. The overlap of the geographic distribution of the attributive form of the verb and the distribution of the folk culture of the *ashagi* architecture suggests the presence of some kind of dialect or culture zones.

Language and folk culture are both closely related to the people in the community. Dialect zones and folk culture zones overlap in some areas, while the do not match in other areas. What is the cause of the overlap or the mismatch? Is it influenced by migration of people or language contacts? How far back in time can we trace the origins of dialect and folk culture zones? These interesting questions are yet to be answered.

The Possibilities of Linguistic Phylogeography in Dialects

If the distribution of the phenomena of the Ryukyuan languages and the variations of individual dialect systems can be mapped out on a linguistic phylogenetic tree and linguistic map, so that the linguistic changes can be plotted on three-dimensional coordinates of time and space, it should be possible to elucidate the process of peoples' migration to the Ryukyu Archipelago, divergence of languages, and the history of how the language developed through human and cultural interactions between Kyushu and the Ryukyu Archipelago or within the islands of the Ryukyu Archipelago.

Ryukyuan language studies, sharing research findings of adjacent disciplines including History, Archeology, Physical Anthropology, Folklore and Folk Music, are essential to understand the developmental history of the Ryukyu Archipelago, including human migration, cultural propagation, and community formation. This kind of integrated interdisciplinary research is called "Linguistic Phylogeography in Dialects", which is interdisciplinary research jointly conducted by researchers in adjacent sciences such as History, Archeology, Physical Anthropology, Folklore and Folk Music.

Acknowledgments This paper is a part of the research supported by a JSPS Grant-in-Aid for Scientific Research 17H06115 (Scientific Research (S), "Comparative and Historical Linguistic Study of the Ryukyuan Languages using Linguistic Phylogenetic Trees") and the Dean Leadership Project of the University of the Ryukyus, "Seeking to Constructa "Dynamic" Linguistic Phylogenetic Tree System for Ryukyuan Languages".

References

Chiyo, K., & Takahashi, T. (2004). *Yorontohougenjiten* (p. 801). Musashino Shoin.

Hashiguchi, M. (2004). Kagoshima hogendaijiten (p. 1011). Taki Shobo.
Ikema, S. (1998). *Yonagunikotobajiten* (p. 381). Jihishupan.
Ishizaki, K. (forthcoming). *Amamitatsugohogenjiten*.
Izenajimahougenjitenhenshuiinkai, (Eds.). (2004). *Izenajimahogenjiten* (p. 1031). Izenasonkyoikuiinkai.
Karimata, S. (2006). *Ryukyushogo no asupecto/tensutaikei no keishiki*. In *Ryukyushogo to kodainihongo: nichiryuusogo no saikennimukete* (pp. 121–143). Kuroshio Shupan.
Karimata, S. (2018). The linguistic difference between northern and southern Ryukyuan from the perspective of human movement. *International Journal of Okinawan Studies, 9*, 15–26.
Karimata, S. (2019). Influence of linguistic contacts and differences between northern and southern, Ryukyuan Language's. *Studies in Dialects, 5*, 5–23. Dialect in Japan.
Kokuritsu, K. K. (Ed.). (1963). *Okinawagojiten* (p. 854). Okurasho Insatsukyoku.
Maeara, T. (2010). *Taketomihogenjiten* (p. 1562). Nanzansha.
Miyagi, S. (2003). *Ishigaki hogenjiten* (p. 1935). Okianwa Times Sha.
Nagata, S., & Suyama, N. (1977). *Amamihogenbunruijitenjoukan* (p. 1013). Kasama Shoin.
Nagata, S., Suyama, N., & Fujii, M. (1989). *Amamihogenbunruijitengekan* (p. 973). Kasama Shoin.
Nagoshishihensaniinkai. (Ed.). (2006). *Nagoshishi Honpen 10. Gengo-Yanbaru no hogen* (p. 634). Nagoshiyakusho.
Nakasone, S. (1937). *kagyohenkaku 'kuru' no kunigamihogen no katsuyonitsuite*. In *Nantoronso— Iha Fuyu sensei kanreki kinenronshu*. Eds. Iha sensei kinenronbunnshuhenshuiinkai (pp. 243–258). Okinawaniposha.
Nakasone, S. (1983). *Okinawa nakijinhougenjiten* (p. 885). Kadokawa Shoten.
Oshio, M. (1999). *Okinawa iejimahogenjiten* (p. 699). Okinawagakukenkyusho.
Toyohama, S. (2013). *Miyako Irabuhogenjiten* (p. 1126). Okinawa Times Sha.
Uemura, Y. (2001). Taneg Ashimahogenjiten (p. 375). Musashinoshoin.

Panel Discussions

Prospects for Critical Island Studies

Ayano Ginoza, Ronni Alexander, Elizabeth DeLoughrey, James E. Randall, So Hatano, and Shigehisa Karimata

MC: Alberto Kiyoshi Sakai, Associate Professor in Faculty of Global and Regional Studies, University of the Ryukyus

Opening remarks: Yoko Fujita, Professor and Director of the Research Institute for Islands and Sustainability, University of the Ryukyus

Concluding remarks: Mutsumi Nishida, Vice President, University of the Ryukyus

Panelists:

Ronni Alexander, Professor in Graduate School of International Cooperation Studies, Kobe University

---

A. Ginoza (✉) · S. Karimata
Research Institute for Islands and Sustainability, University of the Ryukyus, Nishihara, Japan
e-mail: ginoza@eve.u-ryukyu.ac.jp

S. Karimata
e-mail: karimata@ll.u-ryukyu.ac.jp

R. Alexander
Graduate School of International Cooperation Studies, Kobe University, Kobe, Japan
e-mail: alexroni@kobe-u.ac.jp

E. DeLoughrey
Department of English, University of California, Los Angeles, USA
e-mail: deloughrey@humnet.ucla.edu

J. E. Randall
Faculty of Arts, University of Prince Edward Island, Charlottetown, Canada
e-mail: jarandall@upei.ca

S. Hatano
Faculty of Tourism Sciences and Industrial Management, University of the Ryukyus, Nishihara, Japan
e-mail: sohatano@tm.u-ryukyu.ac.jp

© Springer Nature Singapore Pte Ltd. 2020
A. Ginoza (ed.), *The Challenges of Island Studies*,
https://doi.org/10.1007/978-981-15-6288-4_7

Elizabeth DeLoughrey, Professor in English and the Institute of the Environment and Sustainability, University of California, Los Angeles

James Randall, the Chair of the Executive Committee of the Institute of Island Studies and Professor of Island Studies at the University of Prince Edward Island

So Hatano, Professor of tourism, University of the Ryukyus

Shigehisa Karimata, Professor of linguistics, University of the Ryukyus

Facilitator: Ayano Ginoza, Associate Professor in the Research Institute for Islands and Sustainability, University of the Ryukyus

Sakai: Let's start the latter part and first we will have a speech by Professor Fujita.

Fujita: Thank you very much and good afternoon once again. My name is Yoko Fujita, and I'm from the Research Institute for Islands and Sustainability. It's been a long day, and you must be tired, but we are going to have very important discussions. I welcome your input and ideas. First of all, I would like to talk about the Research Institute for Islands and Sustainability (RIIS) and the project that is the topic for this symposium. In April 2009, research on Okinawa became more robust and there was also an increasing demand for establishing a hub for the research; in addition, the Center for Okinawan Studies was established at University of Hawai'i, Manoa, which is a very important partner, so University of the Ryukyus established the International Institute for Okinawan Studies. At that time, Okinawan studies was becoming very active and robust, and thereafter, it started including political science, international relations, economics, sociology, gender issues, and so on. On the other hand, in order to expand the scope of our studies to other islands, and expand our academic collaboration with other island researchers/institutes both in and outside of Japan, we renamed our institution the Research Institute for Islands and Sustainability.

The project upon which this symposium based is titled "Establishing Regional Science for Islands." Thus, the research challenge for establishing a disciplinary field of critical island studies is going to be studies centered on small islands because islands have historically been considered as peripheral or frontier. Now, however, they are taking on more important roles even for the benefit of non-island countries and international society, so islands should not be overwhelmed by major countries but should rather be autonomous and self-decision based. Actually, the sustainable development of small islands has become a very important agenda, even for major countries or industrially developed countries in terms of access to ocean resources, preserving cultural diversity and biodiversity, and addressing global warming. The academic aim of this project is to analyze the current situation of small islands in a scientific way. In order to explore the challenges and the solutions for sustainable development of small islands, we set four topics for our research. The first is "external relations" which includes research on oceanic policies, relations between islands and continents, and inter-island networks. The second one is research on "island economic systems," which focuses on sustainable economic development. The third is research on "island communities," which aims at establishing ways for human resource involvement and community building. The fourth is research on "diversity," which studies ways to conserve and utilize the abundant diversities of the islands. Scholars who are involved in this project are from various fields, such as history, political science, international relations, economy, finance, sociology, anthropology, folklore, archeology, literature, and linguistics. Project members have been trying to understand "islandness" and also struggling to formulize a regional science for small islands. As one of the achievements of our efforts, we created a liberal arts class titled "Introduction to regional science for small islands" at the University of the Ryukyus. As a first step, we can utilize the advantage of Okinawa as a field for island studies because Okinawa is comprised of 160 islands, including 47 inhabited islands, blessed with various characteristics and rich languages, cultures, social and natural diversity and specificity. Of course, words like "small," "remote," "isolated," or "vulnerable" are often used when talking about islands; however, words like "global," "diverse," or "unique" can also be used to talk about islands' features. The former set of words expresses the disadvantages of islands, but the latter set of words expresses the advantages of islands, and this is a very important perspective to develop in island studies, which is a problem-solving style. We can vitalize our comparative research or joint research with other islands by utilizing Okinawa's advantages. The basis for this joint or comparative research is "sympathy." Of course, there are many commonalities between different islands, but there are many differences as well, so that makes us different from area studies that focus on Asia or the United States. It is a very interesting challenge to develop comparative research based on the commonalities or differences among nations or islands, and it will lead to deeper understanding of islandness and also the need for various interpretations. In conclusion, I would like to present one idea for a framework for promoting island studies that will contribute to the autonomous development of small islands. It is very important that academic theory and social practice give feedback to each other; therefore, we are going to address issues from three different approaches: "normative science," "empirical science," and "practical science." Normative science will provide theoretical study, and empirical science will include simulations or case studies. Practical science will implement the outcomes of these studies in real societies in order to create correlations and collaborations among these three. We need to have a good communication method. That is not easy, but we would like to continue working on this.

Now, we are going to have a discussion of these approaches with the five panelists from the previous presentations. We also welcome comments from the floor. Thank you very much for your attention.

Sakai: Professor Fujita, thank you very much. Now, it's time for the discussions, so first I would like to invite the panelists and Dr. Ginoza, the facilitator, to sit at the front of the room. I would like to request that the audience move to this area, which is kind of vacant. Now, from here, I would like to give the microphone to Dr. Ginoza, the facilitator for this discussion.

Ginoza: Hello. I'm Ayano Ginoza, an associate professor at the Research Institute for Island Studies at the University of the Ryukyus. First of all, I would like to thank the panelists for traveling from near and far to join us for our discussion session. Since we didn't have time for questions and answers for each other's presentation, I would like to take some time to receive comments from the panelists on other members' presentations. Since we have a simultaneous translator with us, we can speak either in English or Japanese. I also welcome any comments from the audience as well.

Hoshino: I am Eiichi Hoshino, a professor of International Relations at the University of the Ryukyus. Since I wrote a chapter on the human securities and also post-colonial complex in Okinawa, so naturally I have a question to Ronni. First of all, you were talking about the human securities, but you didn't use that word. Probably, that's because you like to emphasize the being safe and feeling safe part of that. First question is why? The second one is the relation between colonization of bodies, minds, and emotions, and also the colonization of the institution, the system. You emphasized on bodies, minds, and emotions, but there should be a relation between them. If you have something to say about that, I would like to hear. The third one is the relation between colonization and militarization. If you compare Okinawa and Guam, it's good, I mean both of them are kind of hand in hand, but if you compare like, for example, Palau and Okinawa, maybe it may not. Right? So, you could argue about the colonization as it is or separately from the militarization, not so much with the militarization, so the third question is do you agree with that?

Alexander: Thank you. Those are all excellent questions. The first one is the easiest. So, I did not use human security because I think human security is a trap. It is a trap that makes us say military security should have a human face, but in the end, it is really talking about nation states protecting their populations. And, yes, I think that freedom from fear and freedom from want ought to be emphasizing aspects of everyday life, but the way it is playing out in international politics is that those are being emphasized through traditional modes of securitization. So that is why I did not talk about human security. And, also as you know, because I like animals. Human society and islands and everything else too are so dependent on the creatures that live around us, but sometimes human security does not really think about that. I guess I would answer your other three questions by saying first that Guam, and in some ways Okinawa and not so much Palau, but Guam was definitely colonized for military purposes, so it is a military colony from the beginning. In places that are thought by the colonizers to have other benefits, whether economic, resources, or whatever, then I think colonization happens not in a more gentle way, not in a better way, but in a different way. So, yes, Guam is highly militarized. The same can be said for Okinawa. Colonization, of course, changes institutions, changes the understandings of the way governance should happen. For example, take the idea of marriage. If a colonial power begins requiring people to marry, that represents a change that happens on their bodies and on their personal relations but also on the institution of how we govern our society. You are absolutely right; they do go hand-to-hand. I think militarization is similar. One of the excuses or the legitimating ideas for colonialism is militarism. That understanding of military conquest, of course, is held by the side doing the conquest not the side being conquested, the conquestees, if you will. One of the legitimating aspects of that is that we need this territory for protection, we need this territory for economics, we need it for whatever, and it is done in a military way, or at least in a way that may be disguised but relies on the military and military ideas. The more those military tropes and ideas are used to convince people, to control and govern them, to establish institutions, and the more the established institutions are militarized, then the more colonization becomes entrenched too. I think that they are different processes, but they are very much related. The gendered aspect of that is, I think, very important, too. The understandings of masculinities that are used in thinking about militaries, in getting support for the military, and in the everyday practices of militarization are also gendered practices and so we need to think about all of this together, even though we are not necessarily talking about individual men and women. Individual men and women are not necessarily self-identifying as men and women. Yes, Okinawa and Palau are different. I would have to think more about ways that they are different, but while the colonial process and the project may be different, there are similarities, too. Did you know that people from Palau, although it is independent, enlist and serve in the U.S. military? There is a lot going on there that we cannot see, but militarization of bodies and minds is definitely present. Probably Elizabeth in her discussion of militarization of the Pacific, has something to say about Palau too, even though Palau is also, of course, at the forefront of the sustainability discussion.

DeLoughrey: I wanted to follow up on the idea of the internalization of colonial thought. This was an important conversation that Ayano and I had with my UCLA colleagues Keith Camacho and Victor Bascara in a militarization workshop at UCLA that was published in an online Australian journal called *Intersections: Gender and Sexuality in Asia and the Pacific*. In that special issue of the journal we called for a "critical militarisation studies," which included essays from Teresia Teaiwa, Setsu Shigematsu, Wesley Uenten, and others. One of our contributors, Theresa Suarez, reflected quite movingly on her Philippine military family, including a history of abuse that was enabled by military and patriarchal structures. Importantly, she demonstrated how the family itself was understood to be a kind of military institution. To return to one of the questions—we see that the military changes gender structures and familial structures. How then do you determine how people feel safe in those structures? How can they communicate that? Those are concerns articulated from a humanities' perspective. Ronni, your presentation pointed to the importance of getting these stories out—and not just reading them in terms of numbers but reading them in terms of the kinds of stories they tell. But if we reflect back on what Suarez has argued, we have to raise the question as to how a person feels safe in responding to a questionnaire when they understand that the military is part of the family. How do you then disentangle the internalization of the military within family structure, within Chamorro family structure? How can we help create spaces where participants can articulate their feelings about safety?

Alexander: I think that is part of not interpreting the data so much as interpreting the context in which that question is asked and answered is one response. Another is, for example, from Fukushima. If you go to Fukushima Prefecture now, you will see all kinds of signs and information and discussion about how it is not really possible to be safe, so we will make you feel safe. The way you can feel safe is to, you know, not really think about whether this is safe, or this is not safe, and to believe it is safe when you are told it is. Of course there are also people problematizing this. I think what I am trying to say is that yes, it is important to feel safe, but also that obviously we need to be critical about what it means to say that and what impact it might have.

Motomura: Thank you very much for your lectures. I'm Makoto Motomura from the University of the Ryukyus. I'm specialized in child welfare, but specially for the past couple of years, I've involved in the island study, especially in the community forming. I'd like to ask Dr. Randall about Okinawa and Palau, and the comparison between the two. The population of Okinawa is 1.2 million, and the Palau and Guam populations are much smaller than that of Okinawa. So, when we talk about islands, depending on the size of the population, I think the uniqueness would be different. For example, Japan and the U.K. are also islands, so in the current island studies, how do you manage the population size of each island, and what's your view about population? I also would like to ask the same question to the other speakers as well. Thank you.

Randall: Many of us have been using the adjective "small" to describe the context of our research or teaching, or our institutional structures, e.g., the International Small Islands Studies Association or ISISA. This adjective has been highly contested in the literature, and I have mixed feelings about using the word small. In some of the writing I've done, I say it doesn't matter, and that sort of goes against the grain. In fact, there are some researchers who try to analyze what constitutes a population threshold point or cutoff that would constitute legitimate island studies research. For some, the large islands of Japan may not technically be small islands. The same might be said of Great Britain because, of course, it's too large and therefore it can't be considered to have the same kinds of issues, challenges, and problems as other smaller islands. I guess I don't like to think of it in those terms. I would like to think of it in terms of the concept of islandness, because to me the number of people living on an island is less important than the islanders' individual and collective state of mind. How do people feel about themselves as individuals? What is their relationship to the landscape around them; the social landscape, the physical landscape? For example, if I was living on Manhattan in New York City, which is an island connected by approximately 27 bridges and tunnels would I feel like I was on an island? Would I think of myself as an islander? Then compare this to someone who lives on Greenland. This is of course a much larger island as measured by area but a much smaller population. What would be my state of mind? How would I perceive myself in relationship to the physical and social environment around me? Therefore, the number of people that happen to live in a certain place is less relevant to how people feel about themselves, but also how group decisions are made and how group behavior emerges. Is their decision-making influenced by a collective sense of islandness; that is in turn based upon a set of shared characteristics? For instance, there may be fewer institutions on small islands and there is a greater degree of overlap or heterogeneity in the kinds of things that people do. These kinds of characteristics may lead to some peculiar kinds of decisions or decision-making processes. It's almost the social and political psychology dimension that distinguishes small islands from larger spaces, more mainland larger communities. That's a long answer to your question about the use of the adjective "small" to describe island studies. I'm still uncertain about when to use the adjective small. Of course, as you know, within Oceania, many people are now turning this discussion upside down and using terms like "large ocean studies," recognizing that it's not this tiny little place or tiny little community but rather to emphasize the connectedness and the use of the terrestrial and marine space around them that is important in defining themselves as islanders.

DeLoughrey: I wanted to echo what Jim is saying here. There have been critiques about the use of the terms "smallness" and also "vulnerability" in island studies. Ilan Kelman has a number of articles examining these terms. Some have been concerned that the term vulnerability is demoralizing because it downplays the resilience of islands and the resilience of islanders. I wrote about this in the introduction to my first book, *Routes and Roots*, that when 19th century Great Britain presented itself as an island it invoked a discourse of vulnerability, but this suppressed the fact that it was a global empire. (Not to mention it suppresses the existence of Wales and Ireland and Scotland!) I wanted to point out that the discourse of vulnerability can also be a deflection of the history of empire. To return to Jim's point about the influence of Epeli Hau'ofa, he was the Tongan anthropologist who lived in Fiji for many years and established at the University of the South Pacific an islands study center that he called the Oceania Centre for the Arts. Famously, he wrote in 1993 that it was bewildering for his students to hear small islands characterized as marginal, as vulnerable, etc. so he argued for an epistemological shift from "islands in a far sea" to a "sea of islands." The term Oceania is now used to describe the Pacific Islands due to his important intervention. That intervention was a collaborative one produced at the USP Laucala campus. He first circulated that essay amongst his colleagues, who wrote responses about the strengths and weaknesses in understanding the Pacific as Oceania, particularly in the wake of the military coups which had balkanized ethnic relations in Fiji. Hau'ofa's essay was then reprinted on its own in a collection published by Duke University Press, and since that is more widely available the collaborative component of the conversation has dropped out of the conversation. The final point I'll make is that in the essay he describes watching Pele, the Hawaiian volcano, expand into the ocean, helping him realize that islands are dynamic. He concluded his essay by declaring that "Oceania is vast, Oceania is expanding." He wanted us to realize that geography is a process, it's not a product; the earth and islands are changing all the time.

Alexander: So, I've spent most of my adult life thinking about islands and looking at islands and going to islands and liking islands. I understand, on a sort of emotional level really, that there is something that probably we can call islandness, but what Jim seems to be referring to is actually small communities that are not necessarily surrounded by oceans, but share a sort of collective sense of "we-ness" or understanding of themselves as a collective. I am certain it happens in other places like, perhaps, the tops of mountains, too. One of the things that is so fascinating for me about islands is that they are dynamic and that they very much exceed their physical spaces, so that thinking about islands in only a physical sense seems to even miss the real meaning of how people travel, diasporas, different kinds of connections with the human world and with the non-human world. I am full of questions and I think that relates to our discussions this afternoon about language and how language spreads or the idea that cultural sites can be very different depending on who is looking at them. Those sorts of things speak to the dynamism of understanding islands as moving rather than as more or less stationary. So I question, in some ways, whether, in fact, we could even define what it is to be an island. One last point is about disaster, because I spend a lot of time thinking about, and working in, disaster-struck communities. Vulnerability and resilience are big words in the disaster community and tend again to be conceived of and understood in ways from outside rather than from inside. In Haiti after the earthquake and tsunami, for example, people were said to have been resilient, and as such they were nice and brave and wonderful in the face of trauma. But that got turned into a reason for not giving them aid. So I think we need to also, when we are being critical, to be critical of how we think about vulnerability and how we think about resilience.

Ginoza: I think your point on vulnerability in relation to disasters relates directly to the one in Puerto Rico, a very recent incident, and the U.S. aid to the people of Puerto Rico, which was delayed very much. Professor Karimata, would you like to add your perspective to the conversation?

Karimata: Maybe I have different perspectives. We have small and large islands, but when we say small or large, it's a matter of a comparison, it's in relative terms, maybe small in comparison with Okinawa Island. It is not rare for people to understand they live on small islands. Similarly, it is rare for those who live on larger islands to think of its disadvantages. It's only a matter of how you relate your own island to others, and I'm not saying it's not important to realize the smallness or size of the island. There are also majority and minority relationships within the majority. It is important to realize from what stand point you look at the island.

Randall: I hope you will bear with me for a moment, especially for the translators because knowing that the question on islandness came up earlier, I happen to have a file on my laptop that is a short story by David Weale, a retired History professor at the University of Prince Edward Island (UPEI). Although this was published in 1991, when I read it several years ago, it struck me that this is what islandness might mean to some people. He thinks of islandness in some ways as the link between the physical environment and the social. He writes, "I was driving with my ten-year-old son along the shore on the way to a late afternoon hockey game in a town an hour or so away. He sat quietly in the seat just looking out the window at the passing landscape and seascape. I turned to look at him several times, but he didn't even notice. He was absorbed in his looking. Then it occurred to me what he was doing. He was taking in the landscape. He was, if you will, ingesting the Island. And that is exactly what happens when you live here for long—you take the island inside, deep inside. You become an islander, which is to say, a creature of the Island. Islandness becomes a part of your being—a part as deep as marrow, and as natural and unselfconscious as breathing." He goes on, but to me, that's what struck me as being important about what it means to be an Islander. Even though I'm not an islander, I get it now.

Ginoza: Thank you, Dr. Randall. I think I can relate to that quote about the notion of sense of place. I believe that relates to Professor Hatano's presentation on cultural heritage and how islands have been defined by the colonial gaze. Would you like to provide any comments on islandness or islanderness, drawing from your landscape theory?

Hatano: Thank you. As Dr. Karimata said, the concept of "island" is actually determined in a relative sense, and it was also mentioned in Taiwan, too. I talked about the Jinguashi mines today; however, there were also two other islands, and these were close to mainland China and only one kilo or two kilos away. It was actually the frontline for the conflict between Taiwan and China. Many military facilities were built, and military personnel came, so the islanders who had been living on farms or fisheries changed their occupations to ones aimed towards those military personnel, so they began, like, providing sex services or other services to military personnel. Their lives were actually rich; however, in the 1930s, those people left, the military personnel left, so the islanders who had established businesses such as restaurants or dry cleaning for the military personnel lost their businesses, and they could not make a living. My background is architecture, but I'm actually more interested in people's movement, so in that area, people used to live on farms or had small-scale businesses, but because of the incoming military, their lifestyle changed, and when the military personnel left, their lifestyle also suffered, so in a situation affected by the military or the government, those islanders were the ones who were affected the most. Then, is there anything we can do by thinking about the landscape, which is their living place? We can convert those landscapes to protest or oppose or fight against those authorities or bigger powers, and we can think about how they can survive in that kind of situation, so in that sense, their islandness can be considered a vulnerability, but we can think about how we can turn the vulnerability into something of power for them.

Ginoza: I wonder if Professor Masaaki Gabe from RIIS, who specializes in international relations, can take part in the conversation by addressing a question on how islandness relates to the question of what constitute islands. For instance, do islands need to be populated, or can they be without any human inhabitation yet full of diversity or of importance to the national security? I am thinking of the Senkaku Islands as an example. How would you define islandness in case of the Senkaku Islands?

Gabe: Thank you for giving me the time to speak up a little bit. I'm not quite sure about islandness because I just moved to this institution last April, less than a year ago, and this is not an excuse. I just think, first of all, without using islands, how can you explain the phenomena you are talking about, so that's a big point. For example, you say that people kind of live together, you know, that they have kind of collective bodies, so even on continents they live independently, you know, scattered, so they are kind of individuals. Individuals means, you know, sometimes the same as the islanders. I'm just curious about, as Professor Karimata mentioned earlier, why we Okinawans just call this island, usually this Okinawa Island, "Okinawa Honto," which means the Okinawa main island. Without Okinawa, mainland means Japan proper, but they don't use the Japan mainland from the Japanese viewpoint. When I "discover" islands, they're always small islands, but for example, Singapore is an island state. Others, Samoa or Papua New Guinea, are considered islands, and big islands and small islands, so it's very hard to define what islands are. I'm a political scientist, but say, you know, I live on an island, so I'm just an islander with an islander's viewpoint of political science, but I cannot imagine islander political science. It's very hard. I just focus on the power center and outside power, and power always influences people to do something, you know, to say that what to do, what to think. It's very hard to say that physical and geographical conditions determine how people think, so sometimes we just imagine worldwide, beyond geographical limits, so we can think about things. I think if I say that critical island study should be not limited to the physical location, beyond the physical location, we can see much bigger issues. For example, Elizabeth mentioned earlier this morning the militarization not only of inhabited islands but also non-inhabited islands, so Ginoza-san mentioned, for example, Senkaku Island. No one lives on that island, but we discuss the island from the people's viewpoint, so we just always think about our own eyes, from the birds' viewpoint, that Senkaku is very important, I think to people. It's a little bit like we have to move the viewpoint on the island, so that would be much better. As Hatano-san mentioned, an island is an island, socially, for example; the government changes sometimes, but the island becomes a hot issue, so if China controls every island over here, it's no problem with us, right, so if Taiwan controls the whole of China, no problem with those islands. An island is not only a geographical location, so we just think about our imaginative, imaginary object sometimes.

Ginoza: Liz, I wonder if your arguments of Anthropocene can also help understand the way we think of islandness, nation states, and security.

DeLoughrey: Why don't I back up a little bit, just so I can explain how I got to the Anthropocene. The first book project compared Caribbean and Pacific islands in relation to discourses of diaspora and indigeneity. I discovered that in the Caribbean, it's not just about diaspora discourse (from Africa, Asia and Europe) but it's also about how to "indigenize" one's relationship to the land in the wake of plantation violence. On the Indigenous Pacific side while the discourse was about connection to the land– the mountain is an ancestor and progenitor—there was also a diasporic discourse of arrival by voyaging canoe. The struggle I had in the book was trying to find out how to have a conversation through this mixture of geography and history. I posed questions about how you determine islandness through this engagement with geography and history. I realized the islands I was working on were actually archipelagoes, which is also the case for Okinawa. I developed a theory of archipelagos—the term I used in my 1999 dissertation was "archipclagraphy," meaning writing archipelagos, but it's an awkward word and not easy to pronounce! Now in American studies you are seeing an "archipelagic turn," particularly in relation to U.S. empire. So we can challenge the concept of the isolated island, because that is a colonial construct that emphasized isolation and vulnerability rather than the islander's world. How can you have an island that is supposed to be isolated and yet colonist after colonist keep coming? How can the island be isolated when islanders are travelling all over the world? The theory I adopted in my first book was from the Barbadian poet and historian, Kamau Brathwaite, called "tidalectics." He coined this to call forward the relationship between land and sea as a way to understand the complex relation between geography and history. To me that was a much more dynamic way of thinking about the island. That helped me approach the discourse in the Indigenous Pacific where you have a history of voyaging by Indigenous people, but you are also arriving because of the voyaging canoe that brought you there. This allows you to talk about the local and global, connection to the ocean and the land. Now turning to the Anthropocene, it is human centric, as you point out. How would we think about archipelagos as islands from the perspective of the non-human? Some of the work that's been useful for me has been feminist materialism which draws much from Indigenous studies in thinking about how we, as humans, are actually constituted by bacteria and are multi-species beings. This returns us to animism (and vitalism), which is based on the assumption that we are constituted by these other life forms. To go back to your point, one way of ingesting the island is to look at the window as you pass it or cross it, but another way is to labor on the soil and eat the minerals and food from that soil. That's another way of ingesting the island that demonstrates a lack of boundaries between self and other. Obviously I'm still struggling with that question! Sorry for the long answer, but that's the path I've been taking to find a way through. Thank you.

Hoshino: I know this one doesn't help to clarify the definition of islandness, but I would like to throw out, because I was inspired by the Elizabeth's talking about isolation. From human aspect, isolation for the villagers or islanders who lived in the small island, their world could be very small and isolated, but at the same time their word could be very broad, connected to the other island or maybe the continent. So isolatedness is the word or term or concept brought from somewhere else, it could be, right? We could easily find the isolatedness in the big city, New York or Tokyo, so probably I'm going to use the concept of isolatedness in the future, but it should be used a little bit carefully.

Randall: The other term that hasn't been used here, of course, is the term "insularity." In English, at least, and how it's now used in island studies, insularity is used pejoratively. It is derogatory. Island Studies scholars tend not to use the term because it has a connotation of being closed off or being set aside, being lesser, or being other. Do you agree? Although "islandness" has replaced the term insularity, you still see it used in many of the textbooks that were published more than fifteen years ago. So think about that as well in terms of how we talk about language and how language changes. In another ten or fifteen years will islandness become a contested word? I would like to raise one other point here before this session is over. Remember I said in my talk that I was an academic administrator for many years. As is the case within many organizations, acquiring and maintaining resources within universities is very competitive. So, how do we provide our guest administrator present in this room with the ammunition to justify resources being directed into something called island studies. I have tried to do that at a micro-level as a Coordinator of a graduate degree program. When I meet with my senior academic administrator in order to try to justify our existence and try to increase resources, I try to point out the increase in demand amongst potential students. One of my jobs is to increase the demand so that this can be a field of inquiry, a field of study that deserves more resources, and that deserves more attention. That's a microcosm of a conversation that needs to take place on a global level for an island people that are passionate about islands. I also go back to the point I made about what is the existential crisis in island studies. It reminded me that many universities are now putting in place degree programs in Climate Change at the Masters and Bachelors levels. We know, of course, the existential crisis that has precipitated the emergence of these degree programs. So what is the existential crisis that we can use when I talk to my administrators or when we talk to supranational bodies that would justify more resources flowing into an interdisciplinary field of inquiry called island studies? I haven't had an answer to that question yet, but others will.

Ginoza: I wonder if our vice president, Nishida, would like to comment on that.

Nishida: I'm not sure how clear I can be, but this panel discussion represents islandness or insularity or isolatedness, or I mean, represents the discussion of these concepts, and I'm really impressed, and it made me want to also clarify my thoughts. I can only talk about what's in my mind right now. An island for human beings is defined by the islanders or, of course, from the perspectives of outsiders as well. I'm actually a marine biologist or aqua biologist, so for example, if we look at coral reefs, they exist right outside of an island, you know, they are a part of the island, and the island is surrounded by ocean, and I study coral reefs. We study the colonies or populations in each island and study how much they are connected. I studied fish in lakes. Lakes were connected by rivers in the past, but they were isolated by water, so the process of isolation is reflected by the fish that live in each lake, and the rivers are also islands. They are like islands surrounded by land, so the concept of island is like what was mentioned as a relative state, so what we learn from these discussions or these courses can be used for Okinawa, and in seeking the larger concept of island, my knowledge of science can be actually contributed to the reproduction of the definition of the island. I'm sorry, I don't know how much I can make it clear, but I just said what's on my mind.

Ginoza: Thank you, Vice President Nishida. Dr. Karimata.

Karimata: I heard the comment from Dr. Nishida, who reminded me of the following. One thing is that an island is surrounded by the ocean; this is one condition to call some place an island, like an oasis in the desert or some isolated community in high mountains like the Himalayas. This could be only a metaphor. However, because islands are surrounded by oceans, an island itself has unique problems. When we look at the Earth, all lands are surrounded by oceans, so an island can be defined as a place or land area that has advantages and disadvantages because of the fact that it is surrounded by water. Another important aspect of islands is the question of whom islands belong to, or more importantly, where they belong to. Does an island become public land of the local government if the island's sovereignty belongs to Japan? Whose island would it be if the island was already populated before the sovereignty belonged to a country? Another important consideration would be how many rights the islanders have had and exercised in the management of the island.

Ginoza: Thank you for those very thought provoking comments. At this point, I would like to also remind you of three questions we have prepared for this discussion section. I believe we have already addressed the first one on islandness. The second question is what disciplinary or interdisciplinary theoretical and method ological approaches are useful in developing the field of island studies. Third is what the prospects and challenges are for developing the field as a "disciplinary field" to quote Professor Fujita.

Gabe: I have a kind of question. We were talking about physical location of the island and also talking about the community, composed of human beings. They discuss the dynamics of human migration on different islands. Here in Okinawa, even if we look back several centuries, we just move people back, move from the north, in a sense; before, people moved up here to the north. In long-range history, people are moving around the islands. If we focus on the migration among islands, how do you define the islanders? Once you get on a specific island, do you call them an islander of that specific island, so if they lived on the island, they are not different islanders, or are they different? I'm just talking about islands and the community. How do you relate to the community? For example, culture is created by human beings. The culture will move with human migration. But the culture is influenced by the geographical conditions, so how do you think this kind of dynamic of human moving turns migration? What do you think about this kind of argument?

Randall:It's interesting, because let me start my answer by giving you some labels on different islands and how they perceive themselves in relationship to others. Where I am from in Atlantic Canada, I am referred to as a "come from away." On the island of Nantucket in Cape Cod, sometimes they call people who weren't born there "wash ashores," as if they have washed ashore from the mainland. In the Orkneys, north of Scotland, non-islanders are called "ferry loopers," named after the ferries that take them and back and forth between mainland Scotland and the islands. I think sometimes, in our own minds, we simplify things into binary relationships. You are either one of us or you are the "other." We all know that the world is much more complex than this. And this is one way it plays out between islanders and outsiders. I'm sure Liz and all of you and others, if you look at the example of islanders who live in the Pacific, historically they've been very mobile, but even in modern times, you have this phenomena of islanders who have second homes in New Zealand, but because of issues of kin and culture and land ownership, they move back and forth between New Zealand and their other island homes; so what are they? They don't lose their islandness or their island status just because they happen to live in New Zealand. All that highlights to me is the fact that we often misunderstand and oversimplify the complexity of what constitutes an islander. Although there are some people that say that I don't have a right to speak about islands because I wasn't born on an island, I think that's too limiting. What constitutes islands, islanders, and island experiences is a much more complex set of questions.

DeLoughrey: Thank you. That's a very good provocation. I must go back to Epeli Hau'ofa's essay about our sea of islands. One of the things that was challenged by his colleagues at the University of the South Pacific was an idealized notion that we all belong to the ocean in the same way. When Hau'ofa was writing of this concept, it was on the heels of the military coups in Fiji that disenfranchised the Indo-Fijian population. His utopian vision was aspirational, not a reality. In that sense he argued that we need to think of ourselves as part of the planet, we are part of the sea. He argued that Oceania is a mother; therefore we are islanders in a world ocean. He was trying to get us to think beyond all kinds of racial and ethnic limitations. I believe we need utopian ideas to help us through some of the ugly violence of the day to day. To bring us back around to Jim's provocation—namely what is the crisis that we need to get funding for island studies—I have an idea. If you look at the dialogue about climate change, so much is about the limited resources that we have. We realize we, as a planet, are finite and that we have limits in terms of our resources much in the way that islands are understood as finite. In that sense we have an allegory of the island being like the planet. When you look at the documentaries about Tuvalu, Kiribati, or any of the islands that are so-called "sinking islands," you see this allegory at work. The islands stand for the world, and they employ ecological empathy to generate care about Oceania. The island as a figure for the planet is suddenly critical to the ecological imagination. In this way we can understand ourselves as beings of place but also the planet. It's aspirational, like Hau'ofa's work on Oceania, and perhaps utopian, but that is critical to generating new imaginaries for the future.

Alexander: I liked Gabe-sensei's question. There is a field in political science called border studies that I don't know a lot about, but I dabble sometimes, and I think many of us are talking about something that has a lot to do with border studies. Where do we draw those lines and why do we draw them in and around islands? There are serious issues of power and economics and politics that make us do that, Senkaku included. One methodological question, and this is also in reference a little bit to the disciplinary question, has to do with not only what is islandness and what are islands, but also why are islands seen as borders. In fact, we know they are not, they are much more than that. The other response that I would have to Gabe-sensei's question is about time. How do we understand time in the context of islands as opposed to in the context of other configurations of human settlements? I certainly do not know the answer to that, but I think it is an important thing to think about. Particularly when we have populations that are moving and that we know they are moving in ways that are maybe more visible on islands or more understandable on islands. In response to the question about why should anyone give us money, it is because we are who we are. Maybe it is the crisis and maybe it is the idea that islands make people think about what their tomorrow might be, but I think there is a positive way to think about that too, that the ways that the islands contradict our understandings of dichotomies, the ways that the islands contradict our understanding of borders if you will, and of yes or no. The island does not end where ocean is becomes visible.

Those understandings, I think, are essential for finding a way to get out of our current political system and crisis. We have to find another way to understand the world and relations in our world. Perhaps, because islands are so dynamic and they are so much bigger than their physical spaces, a multi-disciplinary understanding of islands can provide an alternative. Maybe it is utopian, but maybe it is not. Islanders kill each other too, but at least they have a different way of seeing the world and of governance. That, I think, is the real strength of our focus on islands.

Karimata: My answer to the question asked by Dr. Gabe would be that your sense of belonging positions islanders. There are people who have residency and places to live, but they may only visit islands occasionally. Others were born on islands but left for a few decades. Some of them return to the islands because of their sense of belonging to the islands. Thus, I believe a sense of belong is crucial. The Japanese word shima connotates territory. Shima doesn't only mean an island surrounded by ocean but also means a territory clearly distinguished from the rest. A shared understanding of the world among the people within the territory is important. That is to say, those who share that understanding are shimanchu (islanders) while others who don't share it are not shimanchu. It's just the way you feel.

Hatano: The current topic can be discussed by me based on my field. I'm studying Taiwan. In Taiwan, a sense of belonging is not that important because Japanese came to Taiwan, then occupied Taiwan, and some Japanese blended into Taiwan. Some Taiwanese accepted the Japanese, but some Taiwanese didn't really accept them, of course. The movement of people is important, but we need to look at the contact zone where people encounter others, so belonging is important, to where Taiwanese belong is important, but from my experience, the relationship between people should have more focus. Considering the flow of people or movement of people, what happens in a colony is that with the movement of people, the movement of things also happens because when people move, of course, things also move. It's not just people taking things when they move but also the ideas or way of thinking also go, for example, because when Japanese people move, a Japanese-like landscape is created, was created in Taiwan, so moving among islands was mentioned, but people and things cannot be disconnected. These are related, so that can be one approach to looking at islands.

Sakai: I would like to go back to interesting discussion about how island studies can contribute to not only getting funds but also new different perspectives or different ways of thinking contesting hegemonic ideas and so on. Maybe, it's also related to this interesting discussion about identity, but just want to share some thoughts, when talking about completed agree. I remember the conversation I had with professor in university of Canary islands, who pointed out the fact that, for example, the Canary islands has when you see it as a part of Spanish state, it's peripheral and GDPs is not so good and so on, but if you change the perspective and see how it's related geographically with Africa, Latin America, historical bonds and everything, it has a totally different potential and everything, right, but for example, they get funds from the European Union I think that because it is considered as ultra-peripheral region, right. I think this is something that Canary island government itself may be applied, I guess they are happy to receive those funds so I don't know if it's a contradiction but I would like to know your opinion because maybe some elites want to get advantage of these ultra-peripheral situation and while in the academia, in scholarship, some people trying to have more positive to provide more positive vision that are also some islanders that, maybe they would not like to do that. Maybe it's also related to colonialism, well it's a different topic but when we were and you were talking about militarism. When you see, when you listen to the opinion that local people have about the bases or whatever, it clearly depends on particular circumstances of everybody, right, so how can we deal with that, because on one hand, we have this vision that come from outside, this point of view that you are peripheral region or whatever, but on the other hand the islanders who are happy with that so I don't know, I would like to know your opinion about this.

Randall: So, there are island studies scholars and researchers who talk about islands not in terms of their vulnerability and marginality but instead in terms of nimbleness, flexibility and adaptability to changing circumstances, with globalization being one of those changing circumstances. And people also talk about political entrepreneurship, for example, how islanders are much more adept at negotiating favorable terms of aid than the donors. This idea of strategic flexibility of islanders to make the best of circumstances is something that I think we should continue to be aware of. You have to be careful when you speak like this because it implies that we don't have to worry about the future of islands, that islanders are always capable of solving their own problems. That masks the complexity of problems facing islands today. It also may explain why the populations of most semi-autonomous island territories, when asked if they want to be politically independent, instead prefer to maintain their political dependence with their former colonial rulers. This is despite the existence of a Decolonization Committee in the United Nations encouraging independence. The majority of the populations on these islands are either self-aware or have been coopted to believe that it is in their own best interests to try to maintain a semi-autonomous relationship, while at the same time negotiating as many powers and resources as they can. Being from Canada, and although it is not an island, I am very much aware that the province of Quebec does this all the time.

Ginoza: Thank you for addressing the question. I think it is related to our third question, challenges and prospects for "critical island studies", funding, aid, and resources. Those are important questions that I think island studies institutions have.

DeLoughrey: One thing occurred to me as I was listening to the conversations about the institute here and hearing about how different kinds of knowledges are being used seems to relate to disciplinarity. Since I am in the humanities and my training is in literature, it's striking to me that island studies has been consolidated as a social science. I'm somewhat peripheral to island studies because I work on island literature and arts. The humanities have not been part of the conversation to the same extent as social sciences. It's important to find ways to communicate across disciplines. One of the things that inspired me was Ayano's comment about the rise of cultural expression at a moment of social and ecological crisis. In my work I noticed that at a moment when the New Zealand state disenfranchises Māori from their ocean holdings to give out mining leases to oil companies, there was a rise in literature and song to counteract that territorialism. So while the state on the one hand is taking away something, Indigenous island culture on the other hand was articulating the ocean as genealogical heritage, as from where Māori derive their sense of being. Ayano, you made a similar parallel about how in the north of Okinawa there are now songs that are being retrieved about the dugongs at the moment of state militarization and destruction of the ecologies of the reefs. So in order to do thorough island studies work, you have to engage the role of culture as a form of resistance to military and/or colonial destruction. In that sense song (and culture) can be used to assert tradition as an act of sovereignty. You have on the one hand a sovereign entity saying here's what I'm going to do because I'm a state or military, and then you have on the other a sovereignty that arises from ontology, culture, and Indigenous knowledge systems. The humanities are critical to thinking about island studies and to understanding how we tell stories.

Alexander: So now, I have to be a social scientist. Because when power is exerted, there is resistance and you are speaking to resistance. In the language of my field, it is resistance and one of the prospects and challenges for island studies, I think, is to look at resistance and look at the way communities push back and redefine their spaces and redefine their world. Particularly in the sense of islands as being more than accessories to mainlands, and that in changing the focus from big powers and big states to smaller entities that are not necessarily states and to seeing them not as smaller versions of the real thing that is big. It challenges all of those ideas of militarization and of patrilocal relations. It changes the dynamics of social relations and it is very difficult to do, because I think that all of us, even with our different cultural backgrounds, we have all been educated, and when we become scholars, we analyze things in term of dichotomies, in terms of who is stronger, in terms of what is more powerful. So island studies gives us an opportunity to work across disciplinary borders but also within our own fields to decolonize ourselves. That is one of the big potentials, I think. Because I have the mic, I wanted also to give a sort of a personal comment to Sakai-sensei's question. I have lived most of my life as a token, okay, as the only woman and the only foreigner and only whatever and one of the things that I learned early on was to find out the biases of the person to whom I am talking to and to use that information to my own advantage. I think every woman in the room knows how to do that, maybe some of the men too, but so do your Canary Island friends. It is one of the skills that people develop. I do not know if this is a good thing or a bad thing, but it is a survival skill and probably it can manifest itself in very useful ways that island studies might be well served in studying. Thank you.

Karimata: I study Ryukyu languages. One day, I realized that the attractiveness of studying Ryukyu languages is that attractiveness that languages of the Ryukyu islands have. Island diversities are in danger and susceptible to big waves and disappear. Island cultures are fragile. I study phonetics and phonetic grammar while assisting the islanders to continue to use their language that they are currently using. I have also been engaged in the study of unpacking language diversity of islands while cultivating ways in which young generations can continue their island languages with financial support from the Ministry of Education, Culture, Sports, Science and Technology-Japan. Once, Japan as a nation moved toward Westernization and modernization while Okinawa moved towards Japanization, and they were forced to assimilate to the one set of values. Nowadays, we are moving back to the opposite direction where we value diversity. We are in an era where we not only value our differences as a characteristic but also where we consider ways to respect our differences. In order to sustain diversity globally, we need research that values islands. In order to carry that out, we ask for funding from the majority. Once, at a large symposium held in Tokyo, I stated that, "Diversity of island languages and cultures is on the verge of extinction. Currents of political power worldwide caused such a fragile situation for islands. That is the fault of the majority, you. Thus, you are responsible for sustaining diversity." I believe it is necessary to speak out thusly as a given right. It is necessary to demand funding by holding the majority responsible for the critical condition of the islands. Instead of asking for funding, as an islander, it is necessary to demand it.

Randall: Could I speak to two things? One is the methodological question that was raised and the other is the concept of resistance. I've never visited Tasmania, but of course Pete Hay at the University of Tasmania in 2006 called for a more phenomenological approach to island studies, and Lisa Fletcher in 2011 stated that we still have some distance to go to bridge that gap between a humanistic and a social scientific approach to island studies. It reminded me very much of the concept of islands as being closed systems or, and I don't really like this trope, "living laboratories." Tasmania might be a good example of that where you've got these university scholars who are all well respected in island studies, and then you've got Tasmania as a microcosm of resistance against corporations and corporate control of the forest and the land. On Tasmania people and local community groups are fighting back against that model. I think that allows us to better understand how islands at both intellectual and grassroots levels can be agents of change.

Ginoza: Thank you. I think that Professor Hatano wanted to add to that conversation.

Hatano: I already said this in my presentation, but if we talk about critical island study, we need to grow out of binarity, because remoteness or isolatedness has been discussed or had been discussed and it's in comparison with mainland but we need to grow out of this binarity or bichotomy, diversity was actually a key word mentioned before and a word I want to use is the word by American Dr. White, he uses the word, practical past, because the past is usually created or recreated by historians but people in that area are actually living in the area not looking at the history book but experiencing the actual fact, so if we just think island from the binary viewpoint, islanders can be minority islanders can be remote but if we admit or recognize that there are multiple pasts and we can say practical past and if we think about this, then we can be more critical.

Ginoza: I appreciate that you brought us back to the important question of the term "critical." The organizers of the symposium thought of this term to describe what

Professor Fujita spoke about in her earlier opening remarks of this discussion section. The term critical describes what Professor Fujita spoke about in her remarks as "a framework that promotes the island studies that will contribute to the autonomous development of small islands" by developing three different approaches—normative science, empirical science, and practical science. We were thinking in relation to other fields' critical terms such as critical militarism studies and critical oceanic studies mentioned in Liz's paper. As Professor Fujita mentioned elsewhere, RIIS has conducted projects creating what she calls Regional Science for Small Islands, where new methodological approaches are formed. One of the objectives of the projects included providing training in the discipline. Receiving a degree in American Studies and entering the job market, I encountered some scholars trained in traditional disciplines who were skeptical of a solid methodological and theoretical background. I wonder if RIIS's projects are councilors of such issues and considered creating a new disciplinary field which Professor Fujita calls Regional Science for Small Islands.

DeLoughrey: I'm curious about you coming to the term "critical" methods, because we had this conversation with a global ecologies group I was working with at the Rachel Carson Center in Munich. We wanted to differentiate ourselves from the people who are not thinking about dialectics, empire, militarism, and power so we came up with the term "critical ecologies," which we derived from the Frankfurt school (German Marxist-informed philosophers of the inter-war period who were anti-militarist). We wanted to borrow from them a kind of self-consciousness of how power works in relationship to people and places, and we adopted the term "critical militarisation studies" for our *Intersections* special issue, so I was wondering if "critical island studies" was also thinking those terms.

Hoshino: We do know in the political science and international relations, critical security studies. Given the security studies that is nation oriented or state oriented and to secure the nations' power and territory and that's the main stream of the security study, there could be a reason why we have a critical security studies. We would like to address the security of the people or human security, right? So it's kind of strange from that point of view, that I couldn't see what's the main stream of the island studies. Already you are talking about the critical island studies, so what's happening here is probably the same with the question Elizabeth is also wondering about.

Randall: I find it interesting what you are experiencing going for job interviews being asked a question about what is your discipline because I think that sort of speaks to some of the questions that we are talking about today. It reminds me that when we are talking about who is it that we are appealing to, to make island studies more important, it's our colleagues, it's people within the tenure and promotion assessment system. If you are going up for tenure or promotion then you have to prove that you have carved out a niche for yourself within your discipline in order to justify advancement within university. It is the same situation for the assessment of grant applications. The evaluators are your colleagues. Unfortunately, they may say that this isn't real political science, or sociology, or geography. So we strike these interdisciplinary committees who effectively give those researchers who operate in interdisciplinary areas a little bit of money to make them go away. It's not the politicians or the bureaucrats that do this, it's us as a community of scholars who are making those decisions. So when all of those little decisions are being made, we have to convince each other of the importance of thinking about things differently.

Alexander: I will try. Because the conversation today, I think, has illustrated why it is critical and that is partly because it is criticizing what those colleagues think of when they think of islands and it is criticizing for me, at least, the way islands are used in security and in military thinking within the discipline of international relations and I assume within all of your disciplines, too. It is critical not only in trying to make diverse but inclusive relationships among academic disciplines in order to build something new but it is also critical because we do not have the right words so we use island studies or islandness or whatever. But there is actually something out there that does not exist in our vocabulary. So it is about denying the current understandings of the way the world is and trying to create new ones.

DeLoughrey: I'm sympathetic to that response because I had a similar experience when I was on the job market. One of the things that critical island studies would have to do is to offer a few critical methods for thinking about islands. On the one hand I see critical island studies, critical ocean studies, critical militarism studies as ways thinking about the discipline in and of itself, so rather than the flat military histories that just summarize how this general of this battle did this, this armored ship sailed here, critical militarization would examine the military as an entity and as an institution. It would examine how it is created with all gendered logics of power, etcetera. It's actually thinking critically about the concept of militarism itself. Critical ocean studies, a term which a few of us have been using, moves away from what's being called "blue humanities," which is a turn to the ocean, but it's not really thinking about the ocean in terms of power, such as exclusive economic zones and so on. It's not thinking about the oceans as matter or territory in the way you find in critical island studies. It means that you are actually thinking about why your project needs the island to make your argument. It means you have to interrogate why this is an island project as opposed to say, a mountain top project, or a project about a peninsula? Why is the island itself critical for your analysis? To go back to Professor Hatano's point about non-binary thinking and the importance of thinking in relation—there's a theorist from Martinique named Édouard Glissant who has been very important to island studies because he writes about what he calls the "poetics of relation," which I used to structure my first book. Glissant is writing from the perspective of a being an islander in the Caribbean, but he also writes of the region as an archipelago. He argues that the dialectic between inside and outside is vital to the island concept. This is similar work to Brathwaite's concept of "tidalectics," and the two were in conversation at a critical time in Caribbean literary studies. To Brathwaite it's about movement back and forth and as a "tidalectic" rather than "dialectic" that can contain paradoxes rather than synthesis. To him, paradoxes are okay so that you are not trying to move towards an easy synthesis. British anthropologist Jonathan Pugh has a number of articles on relationality, using Glissant's theories in thinking about fishing communities in Barbados and about islands in general. This is the kind of epitome of what islandness is—it's about islands in general. In Samoa, it is called the "va," the space in between the two, so that the relation is the most important thing. Perhaps these are possible methods for thinking of what critical island studies could look like, or perhaps tools we could use in approaching the field.

Ginoza: Thank you for a very engaging and critical conversation. Unfortunately, it is time to conclude this discussion session. I hope we can continue this discussion of island studies in the near future.

Sakai: In closing, I'd like to ask the vice president who is in charge at the research and strategy in the University of the Ryukyus, Dr. Nishida to give us closing remarks. Dr. Nishida, please.

Nishida: Unfortunately, we are running out of time, so it's time for me to give you a closing remark. So, I took a lot of notes while I was listening to the comments. I was using my brain much more than usual, so I'm very excited about it. The University of the Ryukyus, which is one of the national universities, 86 national universities in Japan, goes back 69 years, when the university was first established on this island; it has a unique history. As you know, there was a huge or big land battle on the island of Okinawa during the war, and we also know that many people migrated to other countries, including the U.S.A., from Okinawa. Because of the passion and enthusiasm of those people who were living abroad thinking about Okinawa, the university was established as a higher education institution. It was the first one of its kind established on this island of Okinawa to make a contribution to the development of Okinawa. After that, there were faculties and departments added to the existing ones, and we have seven faculties, including humanities, social and natural sciences, and medicine—we do have a university hospital also—so this is a university which is commonly seen anywhere in the world as a university, so the university has its own obligation or mission to create programs to understand the communities in Okinawa. So, I am one of the heads of this program for this group of people who would try to establish contributions to the community, so we are trying to encourage all the classes, the research, and we have established a research center institute; at least two of them have been established. We have a biosphere research institute, this is one thing. It is a facility open to all of the University for Research Work related to the biosphere, as you may understand from the naming. Okinawa is located in the subtropical climate zone, surrounded by coral reefs. Therefore, mangrove research works are a major subject field of study that many of the researchers are involved in. In addition to that, Okinawa has its unique history and culture and has social issues to overcome. Language may be one of them. So, these types of research need to be conducted, or continue to be conducted, and some of them are limited in scale, but under the circumstances, we are trying to encourage and expand our research framework. The Research Institute for Islands and Sustainability is the name of the second research institute that our university has established, and today's symposium was hosted by RIIS, and I feel fortunate to have been here to listen to the wonderful discussions. Ronni-sensei came from Kobe, but you did cross the ocean, right, so three of you came to Okinawa, all the way from overseas, from your country to attend this symposium, and we have two panelists from the University of the Ryukyus, who joined the symposium, and we had a wonderful facilitator, Associate Professor, Dr. Ginoza, so we talked about islands, what that means and how they can be identified. I think what matters is people. Research needs to be expanded, however. One way or the other, research work would be related to human beings or human activities, so we hope that the collaborations between or among the panelists will continue and also that with the participants here in this venue, you may help to develop certain projects. With this, I conclude my closing remarks. Thank you.

Sakai: Thank you very much. This concludes the RIIS symposium. Thank you very much for your input and your participation. Please give a big round of applause to the panelists, and at the end, I'd like to express my appreciation to the interpreters and also the staff members of the institution. Thank you very much.

CPSIA information can be obtained
at www.ICGtesting.com
Printed in the USA
BVHW012354081220
595246BV00001B/14